Femina Problematis Solvendis—Problem solving Woman

David H. Cropley

Femina Problematis Solvendis—Problem solving Woman

A History of the Creativity of Women

 Springer

David H. Cropley
School of Engineering
University of South Australia
Mawson Lakes, SA, Australia

ISBN 978-981-15-3966-4   ISBN 978-981-15-3967-1   (eBook)
https://doi.org/10.1007/978-981-15-3967-1

This Springer imprint is published by the registered company Springer Nature Singapore Pte Ltd.
The registered company address is: 152 Beach Road, #21-01/04 Gateway East, Singapore 189721, Singapore

Preface

This book is the second in what I hope will develop into a long series of texts exploring creativity in its most tangible form. The preceding, and first, volume: *Homo Problematis Solvendis—Problem Solving Man: A History of Human Creativity*—came about because of my joint interests in creativity and engineering. In that book, I set out to explore a series of inventions across 10 time periods, with a special focus on the characteristics that define the creativity of those innovations.

Even before *Homo Problematis* hit the shelves, and despite my intention (title notwithstanding) to make its focus *humans*, and not exclusively men, I did receive a few expressions of concern. Rather than arguing semantics, I had to admit that most, if not all, of the inventions in *Homo Problematis* were by men. I had already been planning a second volume, with so many interesting inventions available for analysis, and it was an easy decision to focus this new effort on innovations by women. Thus, *Femina Problematis Solvendis* was born.

Once I had made the decision to focus on female inventors, I realised I had one problem to resolve. It is quite difficult to find inventions specifically by women prior to about 1800. Naturally, in the Prehistoric era, there are no written records that identify who invented what, male or female. Even in the early periods of *recorded* human history, there is often little information telling us who invented specific things. For these cases, I had to break out of my own stereotyped thinking and ask the question *why not a woman*? This gave me the freedom to attribute inventions like the wheel, the buckle, and the toothbrush to female inventors.

Even from about 1500, in the middle of the Renaissance period, it was tough to find inventions known to be developed by women. The reason for this was a little different from that associated with older inventions, the details of which might have been lost over the passage of many centuries. It was not until well into the 1800s that women in the USA, for example, could own property. A patent for an invention is, of course, *intellectual* property, so there were not just social, but legal barriers that obstructed female inventors. In many societies, from Classical Rome and

Greece, through to more modern times, women have been denied education, political enfranchisement, and legal independence. All of these factors, we know from modern research, and not just anecdotally, impact a person's capacity for creativity and innovation.

Creativity is founded on knowledge. Denied access to this foundation, many women were at a disadvantage when developing novel solutions to problems? Creativity is sparked by personal qualities such as motivation, courage, optimism, and risk-taking. Faced with legal and social obstacles and barriers, how could women be expected to express their problem-solving prowess? Creativity is built on the ability to generate ideas and is helped by a favourable environment. Lacking support, access to resources, and encouragement, is it surprising that the historical record, until comparatively recently, is almost bare of examples of the creativity and innovation of women? The case of Sybilla Righton Masters, who invented the Corn Mill in 1715 (described later in this book), is an example of a time and society that did not allow women to be acknowledged legally as inventors. In fact, what we see in the inventions catalogued in this book is a remarkable story of creativity and innovation *in spite of* the obstacles that women have faced as inventors.

While it was not my intention, in authoring this book, to make any broader political or ideological statement, it was hard not to notice the bias towards men in the history of inventions. If this book helps to draw some welcome attention to the efforts of female inventors across history, then that can only be a good thing. I hope it also inspires young women to pursue careers in Science, Technology, Engineering, and Maths (i.e. STEM). Sadly, female representation in STEM professions is often still lower than it should be.

For me, it boils down to this: if you need to solve a problem, and you deliberately choose to seek help only from half of the people who could assist you, that would seem like a foolish strategy. I cannot imagine going to an emergency department in a hospital but refusing to be treated by a female doctor. "No, no! I'll just wait until a male doctor is free to treat my heart attack!" Why then, until very recently in the recorded history of anatomically modern humans, have we chosen to virtually ignore the ideas of half of our available pool of problem solvers? Regardless of the reasons, you must admit that this seems pretty stupid.

How much faster might our civilisation have developed, had we made full use of the ingenuity of generations of creative, problem-solving women: our *Femina Problematis Solvendis*? How many people might have been spared the ravages of disease? How many lives might have been enriched, and made safer, more comfortable, and more secure?

Thankfully, over the last century, we are seeing real changes to this state of affairs. Perhaps, it is no surprise that technological progress is faster now than at any time in human history precisely because women are able to take their rightful places in the community of creative scientists, engineers, and problem solvers in all disciplines? If diversity, especially gender diversity, helps organisations to be better problem solvers, then it does not seem a stretch to assume that the same applies to whole societies!

In the twenty-first century, we face some serious problems—climate change and antibiotic-resistant bugs, to name just two—and it is now, more than ever, that we need to deploy all of humankind's creative capacity. The good news is that *Femina Problematis Solvendis* have been quietly leading the way since anatomically modern humans first walked the African plains so many thousands of years ago. We need these *Problem-Solving Women* now more than ever!

Mawson Lakes, Australia   David H. Cropley

Acknowledgements

This book is dedicated to the unknown female problem solvers of history, in particular, the women whose creativity went unrecorded and unacknowledged by our ancestors. To those *Femina Problematis Solvendis*, whose inventions kept us warm, clean, fed, safe, happy, healthy and that allowed us to reach our potential, *Ave*[1]!

In the preceding book to this one—*Homo Problematis Solvendis*—I paid a special acknowledgement to humankind's capacity for creativity in the field of medicine and health care. This was prompted by my own experience of being treated for a detached retina while writing that book. While drafting *this* book, I spent a month visiting a university in the city of Brest, in France. After suffering a further problem with my eye while in Brest, I got to try out the French public health system. It is clear that the French, like Australians, place a strong emphasis on high quality, universal health care. As a foreign citizen, I was charged only a nominal fee for four visits to the Morvan Hospital, where I received excellent care from the ophthalmology doctors, the majority of whom were women. It reminded me not only of humankind's ingenuity and creativity in solving medical problems, but also of the many contributions made in this field by women, even as far back as ancient Greece. My favourite, as you might expect, is the *Laserphaco Probe* developed by Dr. Patricia Bath in 1981 (which you can read about in this book). This device ushered in a new era of minimally invasive eye surgery! As I plan to write a third volume in this series, I do hope that I will not have any more eye stories to write about in the acknowledgements!

Finally, I would like to thank my wife and daughter for reading and commenting on the manuscript, and my father and long-time collaborator for proofreading the final draft.

---

[1]In keeping with the Latin vibe, "Ave" is the Latin greeting meaning "Hail!".

Contents

List of Figures

List of Tables

Chapter 1
Introduction

As far back as 1962, the psychologist Jerome Bruner[1] envisaged a future in which artificial intelligence would take on many of the tasks that humans traditionally perform. We are now entering an Age of Digitalisation, often referred to as the Fourth Industrial Revolution, or *Industry* 4.0, in which we are seeing Bruner's prediction come to pass. As computers replace humans in many *routine* knowledge and physical tasks, we must ask ourselves what it is that we, as humans, bring to the table. The good news is that Bruner also told us the answer. *Creativity* is currently enjoying a renewed surge of interest around the World, and much of this being driven by a growing recognition that the so-called *Future of Work* will be one in which creativity is the key skill[2] that we must possess in order to thrive in a world of artificial intelligence, machine learning and so-called cyber-physical systems. Creativity is, or will be, the one thing that we do that machines cannot, and this is why we need to understand our human capacity for finding effective, novel solutions to the relentless stream of problems that we face.

While there is research into what is called *computational* creativity—that is, creativity by computers—the consensus seems to be that computers and AI are a long way from being able to exhibit *true* creativity. There are examples of computers producing, for example, works of art. However, like the computer program *ELIZA* in the 1960s,[3] the question is whether these examples are merely sophisticated mimicry, or genuine examples of creativity. Where the real difference between human and computational creativity lies, in my view, is in the *problems* that drive creativity. In this book, we are interested in creativity as a response to human needs. Those needs define problems, and the problems require novel and effective, i.e. creative, solutions. How could a computer decide, for example, that a woman's corset is too tight, uncomfortable, and unfashionable, and come up with the idea of a brassiere

---

[1] Bruner, J. S. (1962). *On knowing: Essays for the left hand*. Cambridge, MA: Belknap Press.

[2] I prefer to use the term *competency* rather than skill. A competency involves knowledge, attitudes, behaviours and *skills*, so is a better representation of what creativity actually is.

[3] ELIZA was a natural language processing program developed to show the limitations of human–computer communications. Despite this, many users attributed human-like qualities to the program.

© Springer Nature Singapore Pte Ltd. 2020
D. H. Cropley, *Femina Problematis Solvendis—Problem solving Woman*,
https://doi.org/10.1007/978-981-15-3967-1_1

(you can read about this later in the book)? There is no doubt that computers play a vital role in *assisting* in the solution of problems, but for me, the act of *identifying* the problem that needs to be solved will remain a uniquely human concern, at least for a very long time to come. Creativity begins with a well-defined problem driven by human needs and therefore will always be driven by people.

The subject of this book, like its predecessor, is a presentation of the history of the creativity and problem-solving ability of modern humans, through examples of solutions to the range of human needs that have been developed over many centuries. The title, *Femina Problematis Solvendis*, is a twist on the system of scientific classification of human species (e.g. *Homo habilus*, *Homo erectus*), not only suggesting that a defining characteristic of *modern* humans is our fundamental ability to solve problems (the *problematis solvendis* part), but also acknowledging that this ability is not confined to males (hence *Femina*). You will certainly not find my classification in any scientific texts on taxonomy, and it remains intended as a literary device, one designed to help us examine ourselves in a different, and unconventional light.

One critical issue that needs to be addressed, before we begin our journey through the history of creativity, is the question of gender. This is not a problem for most of our inventions. From about 1500 CE onwards, the historical record provides us with unequivocal proof as to who invented what (with occasional exceptions). However, earlier in history, and of course, in prehistory, we run into the unknown. I have selected some common inventions—e.g. the wheel—and will argue that these were invented by women. Some readers may feel this is unjustified and ask, "why have you attributed this invention to a woman?" My general response will be, "why *not*?" Where I do this, I will try to give a plausible reason. If chance favours the prepared mind,[4] then it seems very reasonable to attribute an invention to a woman, if a woman was most likely to have had both the means, and the motive, to find a solution to the problem in question. None of these examples is intended to reinforce any stereotypes, but simply to suggest a plausible *cause and effect* for an invention. If any of my examples are clumsy, then I hope readers will forgive me and accept that my motivation is only to show that, throughout history, men have not had exclusive access to motivated, prepared minds!

Like its predecessor, this book will address the history of human creativity and innovation first by explaining what creativity and innovation are, and why human "needs" act as the stimuli to problem solving (i.e. creativity/innovation). The book will then explore inventions over ten distinct "ages" of human history, beginning with "prehistory", and moving up to the present "Digital Age". Each era of human history will be covered by one chapter, with three key innovations of that era described in each chapter. Unlike other books that discuss and describe inventions and human ingenuity,[5] this book focuses not only on "what" was invented, or "who" did the inventing, but also on "why" it was invented, and "why" it should be considered *creative*. What need did each innovation satisfy, and how have humans, and especially

---

[4]Louis Pasteur's famous observation.

[5]I recommend, for example, Melissa Schilling's book *Quirky* (2018) published by Public Affairs (New York), or Amina Khan's *Adapt* (2017) published by Atlantic Books (London).

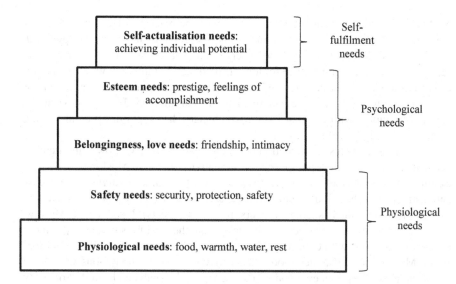

Fig. 1.1 Maslow's hierarchy of needs

women, drawn on their innate problem-solving ability—their capacity for creativity/innovation—to satisfy these needs? In this manner, the book is a history of the psychological ability of humans to find and solve problems (creativity and innovation), and not simply a catalogue of the history of technology, or a biography of female inventors.

The innovations selected for each chapter have been chosen because they are *designed* solutions, i.e. deliberately invented solutions to problems, as opposed to mere accidental discoveries. For this reason, I do not include *fire* as one of the innovations. Although it is impossible for us to know with any real certainty, it seems highly likely that fire was first controlled, probably by one of our evolutionary ancestors, Homo erectus (or Femina erecta?), as much as 1.7 million years ago. *Controlled by* is also important: fire itself would have occurred naturally, due to phenomena such as lightning strikes. However, the ability to create fire intentionally, for example using flints, or fire-sticks, while a purposeful invention, predates our focus on modern humans.

In addition to the criterion of *designed* solutions, the innovations have been chosen because they satisfy human needs such as the need for food, shelter, safety, rest, security, transport, and so forth. We can draw on Abraham Maslow's so-called *hierarchy of needs*[6] as a way of making sense of human creativity and problem solving. Maslow theorised that people are motivated to satisfy certain needs and that these needs/problems have a certain hierarchy, or precedence, to them. Thus, the most basic *problems* that we must solve are those at the base of the pyramid (see Fig. 1.1): how to satisfy our hunger, our need to breathe, our need to keep warm/cool, and so

---

[6]Maslow, A. H. (1943). A theory of human motivation. *Psychological Review, 50*(4), 370.

on. Once those basic needs are met, that is, once the basic problems are solved, as humans we then start to turn our minds to needs/problems higher up the pyramid. If I differ from Maslow, it is only to observe that even the solved problems, or the satisfied needs, rarely stay solved for long. Mother nature rarely allows us the luxury of getting too comfortable. Maslow's theory therefore explains one key driver for human creativity, with *change* explaining another. Our inherent desire to move up the hierarchy, coupled with externally imposed changes, has kept our problem-solving creativity sharply honed for millennia.

In fact, there is a closer connection between creativity and Maslow's hierarchy of needs than may be plain from more recent creativity research. Maslow (1974),[7] for example, discusses creativity in the context of what he calls *self-actualising people* and distinguishes these individuals from the creativity of *special-talent creativeness* (i.e. eminent individuals such as painters, poets, and artists). It seems that Maslow, to some extent at least, regarded self-actualisation, the top of his hierarchy, as more or less synonymous with creativity. Creativity, in other words, IS self-actualisation. While Maslow himself understood that creativity was *not* synonymous with art (a persistent misconception, even today[8]), it seems that he switched his attention away from products altogether and did not explore the idea that creativity could be found in other kinds of outcomes or artefacts. Tying these ideas together, we should understand that creativity is expressed at *every level* in Maslow's hierarchy. At the highest level, creativity is how we solve the problem of self-actualisation, and we can call this *transcendental creativity*. In the middle of the hierarchy, creativity is how we solve the problems of esteem and a desire to belong. We can call this *social creativity*. Finally, at the lowest level of the hierarchy, creativity is how we solve the problems of a need for safety, food, shelter, and so forth. The creativity embodied in solving these basic needs is what we can call *functional creativity*.[9]

Creativity therefore is both the *means*, and the *end*: it is a tool for solving problems, and an end state in itself. In the same way that you don't seek intelligence as a goal in itself (you seek to develop intelligence because it adds value to your life by helping you to do things) so we seek creativity, not merely to label ourselves as creative, but because creativity is a vital component of solving challenging problems and responding to needs.

---

[7]Maslow, A. H. (1974). Creativity in self-actualising people. *Readings in human development: A humanistic approach* (pp. 107–117). In fact, the source of this reference is a lecture Maslow gave in February, 1958—itself interesting in the history of the modern era of creativity. This lecture took place only months after the launch of Sputnik I—the world's first artificial satellite—that event being seen as the spark of the modern interest in creativity.

[8]We only need to look at the cover image of a special Issue of TIME Magazine, published in August 2018, to see that this myth is alive and well. The cover tells us that it is a special issue on the *Science of Creativity* but shows an image of the two hemispheres of the human brain, with the right hemisphere brightly splashed with many colours, while the left hemisphere is blank. Not only does this reinforce the myth that creativity is art, but it also suggests that creativity is confined to the right hemisphere—a notion also debunked in recent years by brain imaging studies.

[9]See, for example, Cropley, D. H., & Cropley, A. J. (2005). Engineering creativity: A systems concept of functional creativity. In J. C. Kaufman & J. Baer (Eds.), *Creativity Across Domains: Faces of the Muse*, (Chapter 10, pp. 169–185). Hillsdale, NJ: Lawrence Erlbaum Associates Inc.

In fact, this end state of *creativity* can be found both in people and in things, and my focus in this book is very much on the latter. This also raises another question. What are the things that can be regarded as creative? In fact, the inventions in this book are not confined to tangible artefacts, i.e. things that you can hold, touch, and feel, such as a shoe. They include *processes* (methods for achieving tangible or intangible outcomes, e.g. a production line in a factory), *systems* (complex interactions of hardware, software and people, e.g. the World Wide Web) and *services* (organised, but usually intangible, systems of labour and material aids designed to meet a need, e.g. education). We will seek examples of the diverse nature of creative outcomes as we explore our catalogue of inventions.

1.1   Ten Ages of Innovation

The different "ages" used in this book were chosen to be representative of the different eras of human development. They are not intended to be a full and complete timeline, nor should they exclude other periods. There is also some fuzziness and overlap, as well as some gaps, meaning that they should be taken more as *signposts* and not strict labels. They are intended simply to give the reader a sense of the character of the period in question, as well as its place in the broader span of human development. Although the ages cover different spans of time, they do give a sense of the accelerating pace of change and innovation as we move forward in time. One note about dates—I will use the designation "BCE" to mean *Before the Common Era*, and "CE" to refer to the *Common Era*. These have the same basic meaning as BC and AD but are religiously neutral. Normal practice is to specify when a date is BCE, but to drop the designation for dates that are in the Common Era. I will use the suffixes for our early time periods, and once we are comfortably into the Common Era, I will drop their use.

The ten ages that we will focus on are;

(a)   Prehistory, covering the period from the dawn of modern humans, up to the invention of writing, around 3000 BCE;
(b)   The Classical Period, spanning the era from approximately 800 BCE to 500 CE;
(c)   The Dark Ages, taking us from about 500 CE up to 1450 CE;
(d)   The Renaissance, spanning the period 1300 CE to 1700 CE;
(e)   The Age of Exploration, covering the years from about 1490 CE to 1799 CE;
(f)   The Age of Enlightenment, which takes us from 1685 CE up to 1815 CE;
(g)   The Romantic Period, roughly the period 1800 CE to 1900 CE;
(h)   The Modern Age, which was essentially the first half of the twentieth century, from 1900 CE to into the 1950s;
(i)   The Space Age, covering a period from the late 1950s into the 1980s;
(j)   The Digital Age, which we can think of as the current era, beginning in the 1980s.

Our focus in this book is on *modern humans*. In scientific terms, we (by which I mean every single one of us alive today) belong to the broader genus *Homo* ("human being" in Latin), and the more particular species *Homo sapiens*. By comparison, the four-legged creature that terrorised Little Red Riding Hood is a member of the genus *Canis* (i.e. dog), and the particular species *Canis lupus* (i.e. wolf). Within our parent genus[10] there have been various species, some of which I have already mentioned. These include the extinct *Homo habilis* (handy/skilful man), and *Homo erectus* (upright man), and also our most recent relatives, *Homo neanderthalensis*. Modern humans, in the sense of Homo sapiens, are thought to have emerged[11] from the evolutionary process as early as 300,000 years ago, probably co-existing, with other species (e.g. Neanderthals) until as late as 30,000 years ago. So-called *anatomically modern humans* (AMH)—us, in other words—have existed more or less as we are now, in both physiological and intellectual terms, from at least 30,000 years ago. The science of this topic continues, and new evidence is constantly appearing that updates our knowledge of the history and evolution of modern humans. However, it seems safe to say that the periods I consider in this book, and the inventions in question, are without a doubt, the work of our direct, modern human ancestors.

Another important consideration is that the inventions I have selected did not necessarily make their *first* (or last) appearance in the epoch that I have placed them in. For example, there is evidence that prehistoric cultures made use of rudimentary toothbrushes however, I introduce the Toothbrush in the Renaissance (specifically in 1498 CE) because this is the approximate date of the introduction of the Toothbrush to Europeans. We should not get too bogged down with the strict definitions of these time periods—my purpose is to capture the essence of different epochs, and not catalogue dates and times. I am an engineer, not a historian!

I ask the reader to remember that our primary focus is to explore the driving need behind the invention and not necessarily a detailed history of the invention itself. For this reason, it is sometimes necessary to be selective in the exact detail of when an invention appeared. Where necessary, I will give some contextualising information, and refer the reader to other sources for a more complete discussion of the invention itself.

In fact, this question of the exact *when* of an invention is consistent with the true nature of *invention*. In other words, there is invention both in an incremental sense—improving on what already exists—and there is invention in a radical sense—the emergence of never-before-seen inventions. In some cases, our inventions will be incremental in nature, and in other cases, radical.

For each innovation, I will follow the same basic format. *What* was invented (a basic summary of the innovation in question); *Why* was it invented (what need did it satisfy, or what problem did it solve); and *How creative* was it (a more objective, systematic scoring of the creativity embodied in the invention). In each case study, I try to summarise the underpinning nature of the innovation by expressing its core

---

[10]Swedish botanist and zoologist Carl Linnaeus (1707–1778) popularised this system of taxonomy, but French botanist Joseph Pitton de Tournefort (1656–1708) is regarded as its inventor.

[11]Another cautionary note here—if you don't believe in evolution, you might want to skip ahead.

purpose or function in the form of "How to verb noun?" In this latter case, I mean, for example, that the basic function of a screwdriver can be expressed as *How to apply torque*.[12] Describing any invention in this way is a good mechanism for understanding not only what it does, but also what other ways might by developed to achieve the same function. Before we start our examination of inventions, the next section explains how I will assess the creativity of the innovations in our catalogue.

---

[12]Torque, very simply, is rotational force.

Chapter 2
Measuring Creativity

A key feature of this book will be the analysis of the creativity of each of the 31 inventions chosen for analysis. This means that our focus is very much on the creativity of the *product* itself, as distinct from the personal qualities, thinking processes or environmental factors that also affect creativity. An important part of the discussion of each invention will be the task of assigning it a creativity score: answering the question *how creative is it*? This will help us to understand and appreciate what problem the invention was trying to solve, and the driving force behind the invention.

Although some variation exists in the body of creativity research, there is a reassuringly high level of agreement about what makes something, *a product*, creative. Most researchers now agree that the creativity of an idea, solution or product must, as a minimum, be defined by:

- Relevance and effectiveness: does the product[1] do what it is supposed to do?
- Novelty: is the product new, original, and surprising?

To explore this topic further, I recommend not only two of my own studies[2] but also the Creative Product Semantic Scale (CPSS) by Besemer and O'Quin.[3] My own research built on these to define additional criteria of:

- Elegance: is the product complete, fully worked out, understandable?
- Genesis: does the product open up new perspectives or change the paradigm?

To show that this scale (the Creative Solution Diagnosis Scale (CSDS), see Appendix A) has a reasonable degree of validity (i.e. does it really measure what

---

[1] I will generally use the word "product" or "artefact" in this book to mean the result of any creative, problem-solving process. This does not mean that the result can only be a tangible thing.

[2] Cropley, D. H., & Kaufman, J. C. (2012). Measuring functional creativity: Non-expert raters and the creative solution diagnosis scale (CSDS). *Journal of Creative Behavior*, 46(2), 119–137. Cropley, D. H., Kaufman, J. C., & Cropley, A. J. (2011). Measuring creativity for innovation management, *Journal of Technology Management & Innovation*, 6(3), 13–30.

[3] Besemer, S. P., & O'Quin, K. (1987). Creative product analysis: Testing a model by developing a judging instrument. In S. G. Isaksen (Ed.), *Frontiers of creativity research: Beyond the basics* (pp. 367–389). Buffalo, NY: Brady.

© Springer Nature Singapore Pte Ltd. 2020

D. H. Cropley, *Femina Problematis Solvendis—Problem solving Woman*,

https://doi.org/10.1007/978-981-15-3967-1_2

we claim it measures) we have conducted several studies that ask people to use the scale to rate the creativity of different products. Using a statistical procedure known as factor analysis, we have been able to show that these criteria hang together in a sensible way, and that they are recognisable in a consistent way by people using the scale. The technique also allowed us to weed out some unhelpful indicators that were not really aiding our attempts to measure creativity and we ended up with a scale that can be used by *anybody* to rate the creativity of *anything*. School teachers have used this scale to give their students feedback on the creativity of assignments, and engineers have used it to rate the creativity or different artefacts. I have even used it to rate the creativity of assorted designs of paper airplanes, to illustrate the different qualities that contribute to a product's creativity.

To help us understand the creativity of the inventions in this book, I have decided to give each one a score, based on the preceding criteria. A product of any sort can be measured on this scale. In other words, we can assign a number for each of the four elements of creativity (effectiveness, novelty, etc.). A typical scheme is to score each criterion on a scale from 0 to 4. Thus, a score of "0" means a complete absence of the element in question. A "0" for either effectiveness or novelty would be fatal for our purposes, and you will not be surprised to find none of these in the book. However, it is possible that we will find products that do not achieve 4/4 for everything. To the best of my ability, I will score the products based on what would have been proper at the time of their invention. Of course, with the passage of time, what may have been highly original 1000 years ago may no longer seem so remarkable to us in the twenty-first century. So, the scores must be considered in context: how creative would the product have been, at the time of its introduction? To help orient our thinking, I will also define the *total* creativity scores (out of 16) as follows: (a) 0–4 is *low* (an invention scoring in this bracket overall is not really creative at all); (b) 5–8 is *medium* (scores in this range indicate an invention that is still rather uncreative); (c) 9–12 is *high* (now we are starting to see inventions that are generally creative in nature), and; (d) 13–16 is *very high* (these are impressively creative inventions)!

In education, there are two contrasting ways of viewing assessment of this type. *Norm-based* assessment looks at individuals in comparison to each other. Under this scheme, a class of students is expected to show a normal distribution of results around some mean value. This means that only a small number of students in the class can be expected to achieve a result far above average, and only a small number far below average, with the majority scoring within a narrower band around the mean value. By contrast, *criterion-based assessment* is concerned with the achievement of set outcomes. Under this scheme, if you show that you can do what is needed, from tying your shoelaces to solving a differential equation, then you pass, and if you cannot show the desired outcome, you fail. One of the features of norm-based assessment that has always concerned me is that it is often used to limit the number of high scores that can be awarded in a class—the idea of *grading to the curve*. This is inherently unfair if a given class has anything other than a normal distribution of ability. Normal distributions are, of course, powerful, and useful, but they are based on the characteristics of populations, or representative samples of populations, and it is easy to see why this assumption might be incorrect in any given class of 28 children.

All of this, of course, is irrelevant to this book, except that I want to explain that I am taking a *criterion-based* approach to the assessment of the creativity of the inventions covered in this book. In other words, I judge each on its own merits, using the CSDS. I do not have a limit of one invention that can receive the maximum score, with the majority required to hover around some mean value. Not only is the list of innovations highly selective, and therefore in no way representative of the set of all things ever invented (in which case it might be possible to make some norm-based analysis), but the basic purpose of this analysis is that if something is creative (if it is relevant and effective, novel, and so on) then it gets a high score, regardless of the creativity of anything else. Having said all that, it will still be interesting to see if there is any regular distribution of my scores, after the fact. The key point is that I am scoring them in a criterion-based sense. I am trying hard to be impartial and objective. Therefore, it is entirely possible that every innovation will have a score of 16/16, or 0/16, or there may be a normal-like distribution of the scores that I give. Let us wait and see!

Another important consideration about the creativity of products (solutions, artefacts, ideas) is that fact that novelty, in particular, is not a static quality. It is probably obvious that the moment an idea or a thing is revealed to the world, its newness begins, so to speak, to wear off. I often give talks about creativity, and the characteristics of creative solutions, and I use a paper airplane problem to illustrate. I start by making a good, old-fashioned, paper dart. I then ask the audience to raise their hands if they have ever seen one of these—the paper dart—before. Everybody raises their hand. I then point out that, if you have seen it before, can it really be considered novel! One of the challenges that businesses face is how to introduce a new product to the market, the iPhone, for example, but at the same time, stop other companies from copying it. It is an impossible task, because the moment you launch the product, you no longer control who knows about it. Not only that, but the effectiveness and relevance of the product seem to be linked to the novelty. Once you launch the product, competitors jump in and bring out rival products, some of which may be better, and the effectiveness of your own product seems to decline. We talk about products becoming obsolete.

So, to a significant extent, effectiveness and novelty are intertwined. The problem is, the only way to maintain novelty is to keep your product a secret, i.e. do not launch it! Obviously, that is a ridiculous proposition, so instead, companies accept this decline, and try, at the very least, to slow the decline of novelty (and the associated decline in effectiveness), for example, by keeping the product secret until immediately before it is launched. Even better, they accept this novelty decline as a fact of life, and work around it by launching a new, better product, fairly quickly, and as soon as the novelty/effectiveness of the original starts to decline.

Possibly, the most interesting, and counter-intuitive example of the relationship between novelty and effectiveness is found in terrorism. This link first gave me an insight into what we called *malevolent* creativity, back in 2005. In the wake of the terrorist outrage of 9/11, I began to hypothesise that creativity was not something exclusively confined to nice people doing pleasant things, like engaging in business. It was clear, from the outset, that the 9/11 terrorists had done something very surprising, and remarkably effective, and I started to explore the connection to the creativity of things.[4] What stood out was that over the space of only an hour or two, the terrorists' *product* (hijacking planes and flying them into buildings) went from highly novel and devastatingly effective, to not quite so novel, and not nearly as effective. The key was that the passengers of United 93, the people who fought back, had found out about the attacks, and decided to act. In creativity terms, the terrorists' novelty had declined to a degree that was sufficient to drag down the effectiveness of their particular attempt to fly a plane into a building. It also illustrated that if the terrorists had thought to take away people's cell phones, the novelty would have remained high, and they probably would have achieved their intended goal.

It is time, however, to return to our core purpose. What are some of the varied problems that our ancestors have faced, across the centuries, and how did they use their capacity for creative problem solving to devise highly novel and effective, and often very elegant and paradigm-breaking solutions to these needs? Let us begin with our prehistoric ancestors and turn first to ancient times to see how and why a prehistoric woman invented the wheel.

---

[4]Cropley, D. H., Kaufman, J., Cropley, A. J. (2008). Malevolent Creativity: A Functional Model of Creativity in Terrorism and Crime, *Creativity Research Journal, Vol. 20, Issue 2* (April), pp. 105-115.

Chapter 3
Prehistory (>2700 BCE): Ancient Invention

Not only has prehistory failed to leave us with written records, it has also failed to leave us with any clear sign of who—male or female—first conceived of many of humankind's earliest inventions. At first, this seemed to pose a problem that could only be solved by making some major assumptions. However, I quickly realised that the very uncertainty of prehistory was also my ally. If we cannot conclusively ascribe an invention from this era to a female inventor, then we also cannot conclusively ascribe it to a *male* inventor! If that is the case, then it is perfectly reasonable to suggest, for example, that the wheel was invented by a woman. We (or at least, I) must remember that there is little justification for assuming that the default must be "a man invented this". In fact, there is growing evidence to support this challenge to the biases and assumptions that are still widespread in society. A scientific paper, published in 2015,[1] has suggested that *inequality* arose only when our ancestors began to shift from a hunter-gatherer lifestyle to one of agriculture. Anatomically modern humans have existed for at least 30,000 years. Agriculture only arose about 12,000 thousand years ago which means that some 60% of our modern human story has been that of the egalitarian hunter-gatherer.

This era takes us from the days of the earliest anatomically modern humans, surviving in small family groups on hostile African plains, through to some early civilisations, such as the Pharaonic Egyptians and the Minoans of Crete and the Aegean. It is tempting to suppose that the needs of greatest relevance in this period were the *basic* ones—how to find food, how to find water, how to keep warm—and yet, archaeological evidence tells us otherwise. Our earliest ancestors buried their dead, for example, often with elaborate care, suggesting that there were also more complex psychological or social needs that drove a search for solutions. Nevertheless, our predecessors could ill-afford to indulge these higher needs for too long, when there were far more pressing matters. Today, a high proportion of the human population can take for granted where their next glass of water, or their next meal,

[1]Dyble, M., Salali, G. D., Chaudhary, N., Page, A., Smith, D., Thompson, J., … & Migliano, A. B. (2015). Sex equality can explain the unique social structure of hunter-gatherer bands. *Science, 348*(6236), 796–798.

© Springer Nature Singapore Pte Ltd. 2020
D. H. Cropley, *Femina Problematis Solvendis—Problem solving Woman*,
https://doi.org/10.1007/978-981-15-3967-1_3

will come from. Thousands of years ago, however, one of our ancestors must have decided that there was an easier way to haul a large animal carcass back to camp, in order to feed the family. Others must have decided that they could travel farther, and hunt for longer, if they could protect their feet from cold, rough ground.

In this chapter, we will begin our study of inventions and creativity by looking at the *wheel*. Once we have examined that, we will then study how one of our ancient mothers found a way to keep her feet warm and protected with *shoes* as she moved around her environment. Finally, as we approach the beginnings of civilisation, we will see how ancient women found a solution to problems of hygiene and health, developing the earliest form of *soap*.

3.1   The Wheel (c3500 BCE)

In the old days, you would chastise people for reinventing the wheel. Now we beg, 'Oh, please, please reinvent the wheel'—Alan Kay, American Computer Scientist (1940–)

Our first invention is a simple example of an energy-handling system. When we think of inventions, in a general sense, there are two that come to mind. Thomas Edison's electric lightbulb, invented in 1879, is universally used to represent the notion of *having an idea*. In *Homo Problematis Solvendis*, I scored Edison's lightbulb at 12.5/16 for creativity (see also Appendix B in this volume). The other invention that is embedded in our subconscious is the *Wheel*. Indeed, it so ingrained in our thinking that we often urge each other to avoid *reinventing the wheel* as though this vital innovation reached perfection more than 5000 years ago! Naturally, it has experienced many incremental improvements over the millennia—changes in the materials used to make it, improvements to its durability, and so forth—and even a few radical changes (see, for example, Mecanum[2] wheels, for a shift in the basic paradigm). However, the simple form of the wheel, and the way that it is used, has remained largely unchanged over the entire lifespan of this rudimentary, but important, invention.

3.1.1   What Was Invented?

Imagine one of our ancestral mothers returning from hunting a large animal with members of her tribe/clan. In Chalcolithic (Copper Age) Europe, this might have been a deer. After the considerable physical effort of tracking and killing a large animal, the prospect of dragging or carrying possibly hundreds of kilograms of carcass must have

---

[2]The Mecanum wheel was invented by Swedish engineer Bengt Erland Ilon and patented in 1972. The Mecanum wheel can be described as a conventional wheel with angled rollers mounted around its circumference. The design makes it possible for the wheel to move in any direction—forwards, backwards, and sideways.

been a daunting prospect, even for hardy hunter-gatherers. Archaeological evidence suggests that far Northern, sub-Arctic humans used sleds, possibly as long ago as 7000 BCE, but these would have offered little help except on snow or ice, and even then, needed many animals to draw along any significant mass. In the time before cultivated crops and the domestication of animals, survival depended on our early ancestors' ability to move: to follow prey, to seek shelter, to find edible plants. We can speculate, however, that our nomadic forebears did not move constantly because it was fun, but because they often had no choice. Some must have faced troubling dilemmas. Perhaps a clan had found a reliable water source near a sheltered and protected piece of land but lacked suitable or abundant prey? Or perhaps the prey was abundant in an area that lacked water or shelter? The tension between the desire to stay in a location with some specific advantages was offset by the deficiency in one key need. What if a way could be found to eliminate the weakness with some clever invention?

It may well have been an early Bronze Age woman, keen to stay where she could safely give birth to and rear her infant, who devised a way to bring far-off prey within reach. From her own experience of hunting she understood the constraints. If the clan's hunters could find an effortless way to carry the fruits of their expeditions over long distances, then a bad location might be transformed into a good one. This is a familiar problem even for twenty-first-century humans. If we had to carry our weekly shopping 20 km, we would probably think more carefully about where we lived. However, thanks to the wheel, the distance of our food source from our home is rarely a factor in our decision making.

So it was that our enterprising *Femina Problematis Solvendis* devised the wheel. The wheel (see Fig. 3.1) appears in the archaeological record with the all-important axle from about the fourth millennium BCE. Wheels joined by an axle appear to have been used for two types of vehicle. The four-wheeled wagon, probably first drawn by people, but later by oxen or donkeys, was the delivery van of this prehistoric period.[3]

Other variants of early wheeled vehicles followed, especially as our forebears transitioned from nomadic to more sedentary, agricultural (and larger) societies. The wagon would not only ease the burden of the hunter-gatherer and offer some relief from the wandering lifestyle but would help trade. Now, humans could bring whatever they needed more easily to themselves: prey, water, grains, building materials, and so on. With this transition also came a second use of the wheel and axle: the chariot. Before the end of the Bronze Age, and typical of humans, this would become an important weapon of war.

---

[3] Image Credit: Wikimedia, Daniel Thornton, Creative Commons 2.0: https://creativecommons.org/licenses/by/2.0/legalcode.

Fig. 3.1 Ljubljana Marshes Wheel, circa 3150 BCE

3.1.2   Why Was the Wheel Invented?

Notwithstanding the later uses to which the wheel, and wheeled vehicles, would be put, the first, rudimentary wooden discs brought about a vital change in the lives of prehistoric humans. With wagons, large loads could be carried across land (see *Homo Problematis Solvendis* for a discussion of the impact of the oar on water-based transport at around the same period) and some of the limitations of nomadic, hunter-gatherer life were eased. If the wheel made it possible for our ancestors to stay in certain locations for longer, perhaps that then gave them the time to notice the potential of cultivated plants as a reliable source of food. All this, quite possibly, because an ancestral woman was driven by a desire to stay in one place so that she could more easily and safely give birth to, and raise, her offspring. The fundamental problem she had to solve to make it possible to stay in one place was *how to move heavy loads*.

3.1.3 How Creative Was the Wheel?

Relevance and Effectiveness: Even though it may have been fashioned from a simple slice of tree trunk, this primitive wheel served the purpose for which it was intended. We all know from experience how much easier it is to move a large stone by rolling it, even if it is not perfectly spherical. Similarly, we have all rolled a log, rather than drag it, precisely because something even roughly circular is so much easier to move. These early wheels worked within the constraints of the available technology to solve the problem of how to move objects. For these reasons, we can confidently give the *prehistoric* wheel the maximum score of 4 for relevance and effectiveness.

Novelty: Two factors help to define how we recognise novelty in solutions to problems. The first is how an invention helps us to understand, or *problematise*, the situation at hand. In other words, how the solution helps us understand what we are trying to achieve. The prehistoric wheel exhibits this aspect of novelty because it offers such a stark contrast to other solutions to the same problem. Take the case of how our ancestors moved loads *before* the wheel. Dragging a heavy mass, even on a cradle (think of the *travois*—a couple of poles, and some netting, used to support and drag a load either by hand, or later, by an animal), must have been a tiresome business. In some terrain, it may have been particularly difficult to drag anything, and the wheel would have resulted in an instantaneous *shock of recognition*: seeing a new solution at once draws our attention to what was deficient with earlier solutions. The second factor that helps us understand and assess novelty is *propulsion*—how the solution sheds new light on the problem at hand. The wheel is essentially a form of *dragging*, but with the problems of friction dramatically reduced[4]! With low friction, we also open up new possibilities for supporting and moving much heavier loads, which would lead to the development of better cradles (e.g. carts) and the introduction of animals to draw these devices. Finally, the wheel would also lead to our early ancestors seeing other uses for this versatile device. For example, placed horizontally, it became a potter's wheel.[5] Taking these two factors together, our prehistoric wheel is not perfect. As new and illuminating as it was as a solution

---

[4]This makes me think of Buzz Lightyear, and his description of flying as *falling, with style*. A wheel is just *dragging, with low friction*.

[5]However, there is considerable debate over which came first: the vertical, load-moving wheel, or the horizontal potter's wheel.

to moving loads, our ancestors must have had a sense of the rollability of some things (stones, logs) so that the novelty of the wheel was more the application of this property, rather than the discovery of the property itself. As I write this, I realise that I am beginning to make an interesting case for the importance of the *axle*, however, let us stick to the wheel: 3.5 out of 4 for novelty.

Elegance: If this characteristic captures the degree to which the solution is well-executed, then the earliest prehistoric wheels, especially those fashioned from slices of tree trunks, might fare rather badly. Of course, as always, we must try to place ourselves in the shoes of a Copper Age human, seeing this for the first time. To our modern eye, the ancient wheel looks rather clumsy, imperfectly round, and not always well-proportioned or well-formed. And yet, the first cart or wheeled cradle fitted with these devices must have been good enough for the idea to catch on, and that must mean that it was sufficiently well-executed to support its effectiveness. We cannot be too harsh on our ancient forebears, and a score of 3 for elegance seems fair.

Genesis: Did the wheel change how our ancestors understood the problem (how to move loads)? Did it open up new perspectives and set a new benchmark for judging solutions to this problem? The wheel certainly showed that there was a dramatically better way to move things. Of course, not only the thing itself—the wheel—but other things as well. The wheel was a new idea in itself and scored highly for novelty. However, it was dramatically new, and not simply a new, improved version of an existing solution. For these reasons, it must score well for genesis as well, and I have given it 3.5 out of 4 in this category.

Total: With a total of 14.0 out of 16, the prehistoric wheel shows *very high* creativity. The major area for improvement, as we know from the way that wheels have developed over the centuries, is the quality of execution. If our ancestral mother had been able to fashion a perfectly spherical wheel, not only would it have worked even better, but this elegance would have propelled it to close to the top of our leader board. It is important, of course, that we remind ourselves that 14 is by no means a bad score. The overwhelming success of the basic wheel concept is evidence of that. It is possible, and indeed it is our task, to nevertheless cast a critical, historical eye on these inventions. If this teaches us one thing, it is that quality of execution is important and adds value to an innovation.

3.2   Shoes (c3400 BCE)

What you wear - and it always starts with your shoes - determines what kind of character you are—Winona Ryder, American Actress (1971–)

The next stop on our journey into the history of *Femina Problematis Solvendis*—Problem Solving Woman—is the humble shoe. As I did in the preceding book (*Homo Problematis Solvendis*), I will allocate each of the inventions that we examine into one of three categories: energy-handling, material-handling, or information-handling systems. The shoe seems somewhat hard to classify in this way: it is clearly not an information-handling system, but is it energy- or material-handling? In the sense that shoes are artefacts designed to help transform chemical energy (in our muscles) into kinetic energy (as we walk or run), then I think we can safely allocate it to the class of energy-handling systems.

One reason for using this classification is that we gain some further insights into the history of human creativity through this system of categorisation. We might tend to assume, for example, that information handling systems are the most modern and therefore the most creative. In *Homo Problematis Solvendis*, we were able to gain some sense of this (in fact, energy-handling systems seemed to be the most creative, by a slim margin) but we will have to wait and see if *Femina Problematis Solvendis* shows a similar pattern. For now, let's take a look at the prehistoric shoe, and the problem that it was trying to solve.

3.2.1 What Was Invented?

Unlike prehistoric artefacts such as the stone hand axe, for which there are many instances in the archaeological record (for obvious reasons: they are hard and durable), being made from perishable materials, the earliest shoes used by our fore-bears have left us with very few physical examples that we can study. That may change in the future, with manufacturers like Nike now making shoes entirely from recycled plastic waste! There is, however, evidence, sometimes circumstantial in nature, of shoes that may go back as far as at least 7000 BCE. For example, by study-ing the bones of our ancestors' feet, archaeologists have suggested that changes in the structure of toes may be the result of the introduction of shoes.

Our knowledge of prehistoric footwear changed, however, with the discovery, in 1991 of *Oetzi*[6]: the well-preserved body of a prehistoric man, found in a glacier near Similaun mountain on the Austrian/Italian border. Prehistory, by definition, has left no written records, so that our knowledge of early modern humans is entirely dependent on the physical traces that our forebears left behind. Among the many fascinating items—clothing, a pouch, a copper axe, and dried fungus—that were discovered with Oetzi were his shoes. Consisting of bear skin soles, deer skin uppers, and webbing made from tree bark, they look (Fig. 3.2) like a pair of slippers. It is not hard to imagine that they provided some level of cushioning and warmth to Oetzi as he wandered across the snow-covered landscape hunting, or perhaps searching for mushrooms or other food, more than 5000 years ago. Whether we believe that in this ancient civilisation men hunted animals, while women gathered edible plants, cooked and raised children, or that they shared these tasks equally, we can readily imagine that it was a woman who first devised these shoes and shared her knowledge

---

[6]In German, it is spelled Ötzi, after the Ötztal Alps, where the body was discovered.

Fig. 3.2 European Chalcolithic Shoe, circa 3100–3400 BCE

with Oetzi. With no written record, there is absolutely no reason to assume that these shoes must have been invented by a man![7]

3.2.2 Why Was the Shoe Invented?

One of the defining characteristics of anatomically modern humans is our status as all-rounders. This may be, in fact, the principal reason why *Homo sapiens* has been so successful as a species. Of course, we have large and complex brains, a capacity for complex speech (and of course, a wonderful capacity for creative problem solving), but it is the fact that we are *quite good* at lots of other things that have also helped our success. Thus, we aren't the fastest or strongest animal on Earth, nor can we fly, or breathe underwater (in our natural state), but we can devise solutions that overcome the deficiencies we have. Once our evolutionary ancestors developed the capacity for bipedalism—walking on two legs—we found that we could exploit this ability even more if we could walk on any surface: hot or cold, rough or smooth. The shoe, from the time of Oetzi, up until the modern age, seems to address the problem: *how to protect feet?*

Before we proceed to analyse the creativity of this invention, keep in mind the reason for defining the problem in the format "how to verb noun?" Creativity is

---

[7]This shoe is a replica of those worn by "Oetzi", the European man found preserved in a glacier in 1991. This replica was developed by Petr Hlaváček and Václav Gřešák, of Zlín, Czech Republic. Image Credit: Wikimedia, Josef Chlachula, Creative Commons 3.0: https://creativecommons.org/licenses/by-sa/3.0/legalcode.

a process of finding novel solutions to problems. Understanding very clearly that a problem exists and being able to define that problem very explicitly (in this four-word form) greatly improves our chances of finding good, creative solutions. Although our ancestors may not have consciously stated the problem in this form, at some level, they must have understood that there was a problem that needed a solution. By attempting to reconstruct the underlying problem, we not only learn something about the task that each of our *Femina Problematis Solvendis* faced, but also have the opportunity to speculate on other possible solutions. Identifying the problem is also a necessary prerequisite to assessing the creativity of the solution. Apart from anything else, the problem statement establishes the baseline against which *effectiveness* can be judged.

3.2.3 How Creative Was the Shoe?

If you have skipped over the earlier sections of this book, in your eagerness to find out more about the featured inventions and their creativity, then I would like to suggest that you pause here and revisit the section on *Measuring Creativity*. This explains the rating system that I use to assess each invention and will flesh out the details of the four criteria that define the creativity of any product, process, system, or service. If you would like to delve a little further, then I also direct you to the appendix describing the Creative Solution Diagnosis Scale (CSDS). The CSDS is the name for the rating system that I use, and the fine detail is in the appendix. If you are ready, then let's get on with assessing, in a more systematic and formal way, the creativity of Oetzi's shoe.

Relevance and Effectiveness: Keep in mind that a score of zero (0) on any of these criteria means, as you would expect, an absence of that quality in the specified invention. Conversely, a score of four (4) will be our maximum. The prehistoric Oetzi shoe not only looks as though it embodies the available technology of Chalcolithic[8] Europe (namely the ability to weave plant and animal fibres into yarn, thread and cordage; and the ability to process animal skins into useable fabrics), but also appears capable of serving its purpose very well. It is not hard to imagine that this shoe would have been reasonably warm, and possibly even quite comfortable, if a little prone to dampness (after all, there was no *Scotchgard* in 3400 BCE). Remember that we will always try to judge the creativity of the selected inventions as if we were seeing them for the first time, in their given time period. I give Oetzi's shoe a near-maximum 3.5 out of 4 for relevance and effectiveness. It may not look like the kind of shoe that we would expect nowadays, with Nikes made solely from recycled plastic bottles, or a *wholecut*, tan Oxford dress shoe, but Oetzi's simple slippers must have been highly valued by their owners for the protection and utility that they offered, especially in a cold climate.

---

[8]Chalcolithic from the Greek "Khalkós" meaning "Copper". In other words, what is called the *Copper Age*.

Novelty: One aspect of this criterion is the extent to which the invention in question helps us understand and define the problem at hand. It is rather curious that one way this is achieved is the way that Oetzi's shoe draws attention to weaknesses in comparable artefacts. In other words, this shoe immediately shows us, or would have shown us, how and why this design is better than simpler alternatives. Whether that was the greater protection it provided to the wearer's foot, or the improved comfort, Oetzi's shoe helps us to understand that the purpose of a shoe is to keep the wearer's foot warm, cushioned, dry and protected from scrapes and cuts. On top of that, novelty asks what new light the invention sheds on the problem at hand. Here, Oetzi's shoe fares a little less well. It is true that it does do a good job of showing how to extend the idea of using animal skins to protect the foot, but it is an incremental improvement to simpler ideas that must have existed previously (perhaps simply binding a rough piece of hide around the foot) and adds only some new ideas to that concept (for example, using cordage woven from natural fibres to hold the shoe on the foot). As a result, this early shoe emerges with a reasonably strong 3 out of 4 for novelty.

Elegance: I have given this prehistoric shoe a score of 3 out of 4 for elegance. Here, we must consider a range of factors that relate to how well-executed, or well-finished, the artefact is, and the extent to which the design is complete, well-proportioned, and fully worked out. In simple terms, we can ask: *Does it look like a good solution to the problem at hand*? Once again, we must resist the temptation to judge this against modern standards. Leaving aside what we know of modern manufacturing methods, and the technology we have available now, I think we can still look at Oetzi's shoe and see it as essentially well made. I know that if I had to make myself a pair of shoes—imagine for a second you are stranded on a desert island—then I would be very pleased to produce something like Oetzi's shoe. It looks like a shoe, and it looks as though it is a pretty good solution to the problem of how to protect feet. There is no doubt that it could have scored better against the criterion of elegance, even at the time, but it is not bad! Therefore, while it is not deserving of the maximum, I think a 3 is justified.

Genesis: In a general sense, genesis asks if the invention in question broke a prevailing paradigm. That frequently happens in one of two ways. Either, an invention is a wholly new and unanticipated solution to the defined problem, or, the solution causes us to redefine the problem itself. In the case of Oetzi's shoe, I think it is fair to say that neither of these conditions is achieved to a spectacular extent. This shoe did not completely redefine the notion of foot protection, and it probably did not cause our early ancestors to redefine the problem. I like to describe this quality as follows: if the artefact in question (i.e. the shoe) makes you realise that "actually, the secret to how to protect feet is ... how to xxxx yyyy" (whatever xxxx and yyyy might be) then you are probably redefining the problem. Oetzi's shoe was a significant improvement on previous solutions, but not a radical departure or redefinition. If we peer more closely into the design, there may be elements that were more radical in their own right—perhaps the cushioning, for example—but overall, its genesis is strong, but not unprecedented. I give it 3 out of 4.

Total: Oetzi's Shoe maintains the strong start to our exploration of *Femina Problematis Solvendis* with a total of 12.5 out of 16. As I did in *Homo Problematis*

Solvendis, I will classify these scores in four basic bands, with score in the range 9–12 considered *high*, while scores above 13 getting the designation of *very high* creativity. The shoe sits on the boundary between these two regions. It is functional, and reasonably well-executed, with elements of novelty and genesis.

3.3 Soap (c2800 BCE)

Soap and water and common sense are the best disinfectants—William Osler, Canadian Physician, Co-founder of Johns Hopkins Hospital (1849–1919)

The last of our prehistoric inventions is a material-handling system. We probably think, perhaps thanks to their portrayal in films, that our ancient forebears were constantly filthy in their animal skins or rudimentary fabrics, and without the modern conveniences that we enjoy today. And yet, we know from archaeological evidence— for example, the location of middens (rubbish dumps) in ancient settlements—that our ancestors had the same dislike as we do for smelly waste. They disposed of it away from their dwellings, presumably because they realised, at least to some degree, that waste was dirty (and unhealthy). Even without the writings of Romans, we know that hygiene had a significant level of importance in that society. Numerous examples of public toilets and baths, for example, in the town of Pompei, can still be seen today. In fact, it makes a great deal of sense that early modern humans would have made proper attempts to keep clean. Without the benefits of modern medicines, it would only have taken one severe, and possibly fatal, bout of diarrhoea to convince our forebears that clean hands reduced the incidence of sickness. It is also very easy to picture, in ancient times, the concern of mothers for some degree of cleanliness and hygiene, given that they probably had the greatest direct role in raising infants. If we think of means and motive, or prepared minds, as a precursor to creative problem solving, then it makes sense that a prehistoric woman would be attentive to anything in her environment that might help to keep herself, and her vulnerable baby, clean and healthy.

3.3.1 What Was Invented?

Soap is believed to date from around 2800 BCE in Mesopotamia. While this just predates the invention of Cuneiform writing (also in Mesopotamia, in around 2700 BCE), our evidence regarding soap is derived from clay tablets found in the region. It is possible, and perhaps quite likely, that soap predated the invention of writing by a greater margin, and these early tablets simply communicated facts and information that already existed.

The recipe that was recorded by these ancient Babylonians combined animal fats with wood ash and water to produce a substance that made cleaning—possibly particularly for cleaning raw wool or cotton fibres, prior to spinning and weaving of fabrics—considerably easier. Raw wool, or *fleece*, is usually dirty, as well as impregnated with oil and lanolin. In fact, raw wool is often referred to as *grease wool* for this reason. Prior to spinning the wool into the yarn used for weaving fabric (see *Homo Problematis Solvendis* for a discussion of the spinning wheel), it therefore needs to be washed. Unfortunately, fats and oils have the unhelpful property that they are hydrophobic, and therefore repel water. In other words, they are not soluble in water. Readers will know, from experience, that you cannot simply rinse butter or animal fats from your hands when cooking: it requires soap!

So, as our ancestors' technological prowess developed—making fabrics from natural and animal fibres, for example—new problems were created in parallel with these advances. Soap may therefore have been developed in response to the need generated by the production of woollen fabric, but the broader applications—helping our ancestors to keep themselves clean—would have followed close behind.

Exactly how soap came to be invented is unclear. Of all our innovations, this may well be an example that actually began as a serendipitous[9] discovery. Our focus has always been on deliberate inventions, and not accidental discoveries, and yet there is a relationship between the two. As Louis Pasteur[10] said *chance favours only the prepared mind*. Even though the first recipe for soap may have been discovered by the accidental mixing of fat, ash and water, it must have taken the prepared mind, as well as the creative determination, of an intelligent, attentive *Femina Problematis Solvendis* to turn these ingredients into a useful, and useable, substance for washing clothing, and ourselves.

In its raw state, as you can guess from the ingredients, soap is not the sweet smelling, perfumed substance that we think of today. Basic, *unenhanced* soap may have been sufficient for degreasing wool fibres, but our ancient forebears were probably no more interested in rubbing themselves with sooty animal grease than we are today. It is only a small additional step to imagine women of the ancient world adding some pleasant-smelling oils to the mixture in order to create soap in a more recognisable form (Fig. 3.3, for example).[11]

Over the centuries, and notwithstanding my comments about hygiene and cleanliness, it is clear that our ancestors have not always understood the relationship between dirt, soap, and good health. Later discoveries, for example, the germ theory of disease, would remind us of the effectiveness of soap as a simple way of guarding

---

[9]At the risk of insulting your intelligence, and because the word is often used in contexts associated with innovation, the origin of *serendipitous* bears repeating here: it is the adjectival form of serendipity. That word was coined by Horace Walpole in 1754, in his story *The Three Princes of Serendip*. In it, three heroes constantly discover by accident things they were not searching for.

[10](1822–1895) and inventor of Pasteurisation, as well as vaccines for Anthrax and Rabies.

[11]Aleppo Soap is a hard soap made from olive oil and lye (sodium or potassium hydroxide, originally obtained from ashes) mixed with laurel oil (a fragrant oil obtained from an aromatic, evergreen tree). Although not the first soap made by humans, there are unverified stories that Aleppo Soap was used by Cleopatra. Image Credit: Wikimedia, Bernard Gagnon, Creative Commons 3.0: https://creativecommons.org/licenses/by-sa/3.0/legalcode.

Fig. 3.3 Aleppo Soap

against many illnesses. We can thank one of our ancestral mothers, her keen eye, and her curiosity, for seeing the potential in such an odd mix of ingredients!

3.3.2 Why Was Soap Invented?

The most likely underpinning problem, for which soap is a solution, as we have discussed already, is how to dissolve fat. Dissolving fat (in water) immediately makes it possible to rinse the offending fat away: dip your woollen fleece in the river, and both the soap, and the dirt are carried away. Solving the problem of how to dissolve fat, however, very quickly opens up a range of sub-problems: how to rinse the fat from the clothing, how to clean skin, how to clean eating utensils, and so on. This would later lead to inventions that could dissolve other undesirable substances, rinsing them away too, until we have modern laundry powders, industrial cleaning agents, as well as antibacterial, non-soap, skin cleaning agents! With our ancestral mother in mind, we'll fix the basic problem as the slightly more specific problem of *how to keep clean*.

3.3.3 How Creative Was Soap?

Relevance and Effectiveness: Even in its crudest, most ancient, and simplest form, soap is remarkably effective in dissolving fat. The chemical mechanism underpinning soap is very simple. Soap molecules consist of two ends, one of which is strongly hydrophilic (attracted to water) and one of which is strongly hydrophobic (repelled by water). When mixed with substances like fat or oil, and water, the hydrophobic ends of the soap molecules attach to the oil/fat, while the hydrophilic ends are attracted to the surrounding water. What this creates is a kind of multi-layered bubble, with fat/oil in the centre, a ring of soap molecules around that, and finally, an outer layer of water. The oil/fat is now, in effect, suspended in water in a protective bubble of soap molecules. Rinsing the oil/fat away is then a simple matter of allowing the water to carry these fat bubbles away.

Having just made the case for the effectiveness of soap, I have to scale back my enthusiasm. Soap, deliberately made, is highly effective. Soap, accidently discovered, may not have been so good. Our forebears may have taken some time to work out exactly how the mix of wood ash, fat and water became soap. This means that once soap had been discovered, it may have taken time to reliably reproduce good soap. The technical explanation would be that soap, to some degree, exceeded our ancestors' knowledge and technical ability. Because of this trial and error, its performance must frequently have been far below what we now expect. What this leads us to is the conclusion that prehistoric soap, deliberately made, began as somewhat less than fully effective. Therefore, it gets a 3 out of 4.

Novelty: In contrast to effectiveness, prehistoric soap must score well for novelty. Nothing comparable existed before this combination of wood ash, animal fat, and water was discovered, but more importantly, harnessed to meet a very real need. Deliberately made soap drew attention to the shortcomings of water alone as a cleaning agent. It showed the potential, with the right technology, for improving methods of cleaning. It took basic concepts of cleaning in a new direction, by recognising that substances added to water could dramatically improve the effectiveness of cleaning. Finally, it also quickly suggested different ways of making use of this substance: for cleaning not just wool, but clothing, utensils, skin, and so on. Soap must score the maximum of 4 for novelty.

Elegance: Because of its initial accidental discovery, and trial-and-error improvement, it is not surprising that the first soap scores relatively poorly for elegance. It would not have appeared convincing to the first observers: how can this smelly mix of what appear to be waste products actually make things cleaner? It was crude, unappealing, incomplete, and certainly not obvious, even if it worked quite well. For these reasons, we cannot score it as highly elegant (even if later forms were elegant), and I have given it a modest 2 out of 4 in this category.

Genesis: Finishing on a strong note, the soap that began as a lucky discovery and was harnessed by our ancient *Femina Problematis Solvendis* was truly groundbreaking. Although it was discovered, rather than invented, this takes nothing from its paradigm-breaking nature. This odd combination of materials stepped off a curve that had long since passed its point of diminishing returns and established a new standard in one simple step. It immediately created a new benchmark for cleaning and would open up a long series of refinements and improvements based on similar concepts. The high score for novelty is complemented by a score of 4 out of 4 for genesis.

Total: Prehistoric soap, first noticed, then harnessed and refined, has achieved a score of 13 out of 16 for creativity. This places it in the *very high* range for creativity. The one cautionary note is that it was not invented, deliberately and consciously, but was a fortuitous discovery. Our focus in this book is invention—the deliberate application of creativity in a process of recognising and solving problems—so that soap might be considered slightly out of scope. However, I have included it because it seems so non-obvious that it must have required the same underpinning skills in problem solving to recognise what was happening and harness the potential of what had been observed. In more modern times, we often have the advantage of a larger body of preexisting knowledge upon which we can draw as we construct creative solutions to problems. In the prehistoric world, this was not the case, and perhaps this makes the discovery/invention of soap *more* impressive?

Chapter 4
The Classical Period (753 BCE to 476 CE): Creative Civilisations

The second epoch in our sequence is the Classical Period of the ancient Greeks and Romans. This period spans about 1200 years, from the eighth century BCE (the founding of Rome took place in 753 BCE) up to the fifth century CE (the fall of the Western Roman Empire occurred in 476 CE). You will notice that there is a gap from the end of prehistory to the start of this period. There were, of course, many inventions in this gap: the Egyptians invented the spoked-wheel chariot and the saw; the Chinese invented the umbrella and inoculation; the Minoans invented the aqueduct; East Africans invented steel. I omitted this pre-Classical Era only because space did not permit.

The Classical Period was a time of great advances in the Western hemisphere. Led by the ancient Greeks, many of the foundations of art, philosophy, democracy, science, and education were first laid down and systematised, before being adopted, refined and spread across Europe and the Middle East by the Romans. Although *lost*, to some degree, after the fall of the Western Roman Empire, it was their *rediscovery* that would be a catalyst for the European Renaissance almost 1000 years later.

The first innovation that we consider in this period is the *Buckle*—the device that you depend on to hold up your pants, keep bags securely closed, and to prevent your sandals from falling off—tackling the simple problem of holding or joining things together. Our second invention in this era is the mysterious *Antikythera Mechanism*, developed in ancient Greece, and a vital advance in humankind's ability to understand and predict the natural world. The final innovation that we consider in the Classical Period is the *bain-marie*. This simple energy-handling system allowed the precise control of temperature needed for what we now regard as the science of chemistry.

The *bain-marie* is also the first invention in our catalogue that we know was developed by a woman. That fact also calls attention to the premise of this book. Even after the advent of written records, it is difficult to find concrete evidence of female inventors. From 300 CE until at least the beginning of the Renaissance in 1300, women more or less disappear from the history of technology and invention. I found a small number of examples that you will read about in due course, but it was a challenge! Our *Femina Problematis Solvendis* only begin to re-emerge, clearly and independently, around the time of the Enlightenment (1685–1815). Even then,

© Springer Nature Singapore Pte Ltd. 2020
D. H. Cropley, *Femina Problematis Solvendis—Problem solving Woman*,
https://doi.org/10.1007/978-981-15-3967-1_4

we sometimes have to dig further and harder, because societal conventions of the eighteenth and nineteenth centuries often denied women the right to own property which, among other things, denied them the right to own patents. Many women worked around this barrier with the cooperation of willing husbands, who often were named as the inventor on patents. For earlier eras, we may never know for sure who, male or female, developed some of humankind's most important inventions.

4.1 The Buckle (c700 BCE)

You might be a redneck if... your belt buckle weighs more than three pounds—Jeff Foxworthy, American Comedian (1958–)

Many of the older inventions that I describe in this book (and in the preceding volume: *Homo Problematis Solvendis*) have evolved considerably over the centuries. The modern oar, for example, is a lightweight, carbon fibre evolution of its ancient ancestor. Modern lithium-ion batteries are only distantly similar to the Leyden Jar of 1745. However, the humble buckle (seen in Fig. 4.1) has hardly changed in nearly 3000 years. Whether used to join the ends of a belt, the strap on a bag, or the chin strap of a helmet (the word "buckle" comes from the Latin "buccula" referring to the so-called *cheek-strap* on a Roman helmet), this device is still widely used today in a form that is almost identical to its predecessor. Of course, there have been incremental improvements (plastic instead of metal, for example) and also more radical changes (think of plastic snap-fit buckles that are also common today). However, the simple

Fig. 4.1 Roman Military
Buckle circa 400 CE

metal buckle shows a durability that may have a lot to do with its properties as a highly creative solution to a well-defined problem.[1]

History has not recorded who invented this device, but I will argue that a woman was responsible. The Latin origin of our modern word suggests that the buckle was first used to secure a helmet more firmly on the head of soldiers in the early years after the founding of Rome. Although Rome was yet to develop into the world power that we associate with generals such as Julius Caesar, or his heir, the Emperor Augustus, the city's rise to the status of military superpower must have been founded on an early superiority over other competing tribes and adversaries. Is it possible that a Roman woman, concerned for her son or husband, found a way to improve his protective equipment with an adjustable, and secure, chin strap? Whatever the motivation, this device quickly showed its utility, not only in military equipment, but for a range of other applications.

4.1.1 What Was Invented?

The buckle pictured in Fig. 4.1 hardly needs an explanation. It is so similar to the modern equivalent that there can hardly be a reader who is *not* familiar with it. Nevertheless, and for the sake of completeness, I'll risk insulting your intelligence with a brief overview of this energy-handling system.

The buckle, in combination with straps, comes in a variety of forms, however the oldest is the so-called *frame-and-prong* style pictured. Consisting of a usually D-shaped metal loop onto which is fitted a movable prong, the buckle operated in one of two circumstances. Either the loop and prong—the buckle—was also attached to a strap, as in the case of a belt, or, the buckle was attached to an object, for example a helmet. In either case, a strap (the loose end of the belt, or a strap attached elsewhere on the helmet) was then fed through the loop, and the prong inserted in a hole in the strap. Pulling on the strap would then draw the prong down onto the loop, preventing any further movement of the strap. In this way, a helmet could be secured on the wearer's head, or a belt fastened tightly around the waist.

The frame-and-prong buckle has the advantage that, provided even a slight tension is maintained on the strap, it is almost impossible for the strap to come free from the prong. In other words, it is very hard for the buckle to undo itself, making it a very reliable way of securing an object. When first introduced by our enterprising *Femina Problematis Solvendis*, it must have quickly made an impact. Imagine trying to tie and untie swollen leather laces on your helmet's chin strap on a cold, wet evening. Think of the advantage of a buckle and strap that could be quickly undone, even with one hand, and yet which would stay securely on your head even in the heat of battle.

---

[1] This copper alloy Roman military buckle was found in Britain, and it is thought to date from the late fourth or early fifth century CE. Image Credit: The Portable Antiquities Scheme/The Trustees of the British Museum, Creative Commons 2.0: https://creativecommons.org/licenses/by/2.0/legalcode.

The longevity of the buckle—some 2700 years—seems testament to the impact that it has had.

Even in ancient times, variations of the buckle emerged. During the Dark Ages, in the eighth century CE, a double-loop design emerged. Think of a standard frame-and-prong, but with two D-loops back to back, and the prong attached to a post in the middle. Other designs were created for particular applications: for example, small buckles with removable centre pins were designed for shoes, while plate-style buckles, with interlocking metal hooks, were common in military forces in the nineteenth century. The latter were designed, in part, to create a flat, outward-facing metal plate that could be decorated with insignia. Nevertheless, even in the twenty-first century, with far more sophisticated press-studs, zippers, Velcro and elastics available, the basic frame-and-loop buckle remains a common feature of belts and is still used on sandals, shoes, and other accessories like bags.

4.1.2   Why Was the Buckle Invented?

The buckle, in 700 BCE as now, solves the problem of *how to secure objects*. If this seems rather vague, and a problem that could be solved by many things other than a belt, then keep in mind that this is exactly the point! We always try to state the verb and the noun in a form that is as abstract, i.e. non-specific, as possible, precisely to encourage the greatest range of possible solutions, of which this solution—the buckle—is just one that happens to be very good. My point is that we are not trying to reverse engineer a very narrow problem statement that only fits this solution. We are trying to show that the buckle was a very good, and therefore very creative, solution to a very broad problem. Some of what makes the buckle so good is the context that we apply, but only after we have explored a wide range of solutions. It is possible that other innovations would also satisfy the problem of *how to secure objects*, but only once we have some solutions to consider, and to which we can apply some constraints. This reflects the essential steps of creative problem solving.

4.1.3   How Creative Was the Buckle?

Relevance and Effectiveness: Keeping in mind that our analysis is of the buckle itself, and not any attachments (so we are not judging the creativity of a belt, for example), I am unable to find any compelling reason to knock points off for relevance and effectiveness. The basic frame-and-loop buckle was *correct*, in the sense that it embodies the knowledge of iron, and the ability to work that metal, that we can expect from early in the Iron Age. It is worth emphasising that the *relevance* aspect of this criterion (as opposed to the *effectiveness* part) seeks to recognise innovations that were not beyond the technology of their time. This is important because it rules out solutions that are unfeasible, technologically impossible fantasy. A creative solution, in other words, must not only do what it is supposed to do, but must be realisable. Therefore, *inventing* a time machine, for example, requires more than just thinking up the concept. It must be feasible, and if it is feasible, then it can be made, tested, and hopefully demonstrate its performance. Clearly, the buckle was not only feasible, but performed its function well. 4 out of 4!

Novelty: Would the buckle have surprised people when first introduced? It must certainly have made the weaknesses of other fastening methods very obvious. The ease with which it could be fastened and unfastened, and its tendency to stay secured, all demonstrated some of the areas where competing methods were deficient. It not only represented a significant extension of preceding methods (e.g. leather ties), but also took a radically new approach to the task of joining objects together. Conversely, it might be argued that it still relied on an underlying idea of two straps, for example, being joined, albeit in a new way. It did not, for example, do away with the need for a chin strap on a helmet. It merely found a much better way to secure this strap. There is much to like about the novelty of the buckle, but it is perhaps not a perfect example of this criterion, and therefore I give it a score of 3.5 out of 4.

Elegance: The buckle is relatively easy to assess against the criterion of elegance. For an observer present in 700 BCE, and even today, the very simplicity of the frame-and-loop design appears well-executed and well-finished. The example in Fig. 4.1 does not look as though there is anything further needed (apart from a strip of leather). It is pleasing to the eye, neat, and simple, and the elements fit together in a clear and consistent manner. If it has a slight weakness, then it may be that the buckle could be (and in modern times would be) slightly better finished and better proportioned. These are slight quibbles, and the fact that the basic design has persisted, almost unchanged for nearly 3000 years, means that it deserves a score of 3.5 out of 4.

Genesis: The frame-and-loop buckle changed our ancestors' understanding of the problem of securing objects like helmets and belts. The prevailing paradigm focused on tying fabric or leather strip in knots was replaced by a more sophisticated and functional innovation. The buckle established a new benchmark for judging methods for securing objects and pointed the way, not only to further improvements of the buckle itself, but to other, more radical replacements (e.g. Velcro). The development of the buckle for the purpose of securing the chin straps of Roman soldiers would also lead directly to the application of the buckle for a variety of other *securing* problems (e.g. shoes). Though not outstanding with regard genesis, the buckle probably not only took a prevailing paradigm to its limit, but also partly introduced a new paradigm. For these reasons, I give it 3.5 out of 4.

Total: 15 out of 16. Although it is far too early in this book to declare a winner, the buckle has made a bold, early claim for the prize of most creative invention. We can also compare it to the inventions in *Homo Problematis Solvendis*, and it sits third equal on that list, alongside the crankshaft. Readers will be able to look at the full set of rankings at the end of this book (both for our *Femina* inventions and in comparison with the *Homo* inventions from the companion to this book).

One final note. I find it significant that the buckle shown in Fig. 4.1 has no strap attached. Of course, this is because the leather or fabric that would have been attached has long since rotted away. However, this also emphasises, for me, just how robust this simple invention is. I've had many belts over the years, and the part that usually fails, whatever the material used, is the strap itself. The simple metal buckle seems to last forever. The exception to this may be a type of belt that I only discovered a few years ago. This new *wonder-belt* has a strap made from a woven, stretchy, synthetic material. The concept of a braided fabric belt does not appear to be wholly new, but by making the fabric from an elastic, *synthetic* material, the whole belt now seems to be virtually indestructible. I suspect that in 3000 years, future archaeologists will unearth these items in perfect condition (and perhaps speculate on their creativity)!

4.2   The Antikythera Mechanism (c150 BCE)

New technology is the true friend of full employment; the indispensable ally of progress; and the surest guarantee of prosperity—Margaret Thatcher, UK Prime Minister, 1979–1990 (1925–2013)

One of the challenges of writing this book is the fact that many of the inventions, even of later periods, have no known inventor. Having declared a specific focus on inventions by *women*, it is easy enough to declare that these unattributed innovations must therefore have been created by females. That seems a perfectly reasonable assertion: why shouldn't that be the case? However, it does open up a discussion that has some significance for the field of creativity. We know that for humans to generate novel and effective solutions to problems, a range of factors come into play. These range from the cognitive (thinking) skills required, through to the personal qualities that drive individuals to seek out new ideas and solutions. The environment in which the innovator operates is also a key factor, and these different elements have the added quality that they function as a system. In other words, creativity results from the interaction of *who* we are, *how* we think, and *where* we work. Once we think of creativity in this way, we can circle back to the role of gender in this system. I am talking, in particular, about the interaction between gender and environment.

Attributing inventions to women throughout history is aided by consideration of the social and physical environments present at various times in the past. This requires us to ask not only *why* something was invented—in the sense of the problem it was solving—but also why in terms of *why then,* or *why her?* Why, for example, was the problem of *how to keep clean* of particular relevance for women? Why did this problem drive one of our ancestral mothers to find a solution? Why might that drive have been especially strong and relevant for women? In the case of the invention considered in this section, why was the problem that the Antikythera Mechanism solved of special significance to women, such that they would have been motivated to solve it? Let us consider the invention in more detail.

4.2.1   What Was Invented?

The Antikythera Mechanism was an early example of an *analog* computer. Digital computers, of course, function by performing various arithmetic and logic operations on discrete binary digits (think 1's and 0's). These binary digits take the form of electrical voltages, and everything that the digital computer processes—words from a word processor, mathematical problems, images from a camera, or a person's voice—must first be converted from the continuously varying world of light, sound, and other forms of energy (the analog world), to digital form. Even the instructions that we give the digital computer must first be converted from the language of humans to this digital language (you can read about the invention that makes this possible in the later section on the Compiler). *Analog* computers, as you may now have guessed, differ in that they operate directly on these continuously varying signals and forms of energy.

This may seem a little strange in a world that is now undergoing such a profound *digital transformation*, but we have all experienced analog computers in action, probably without even realising it. If you are a little older, you may well have used a slide rule to solve maths problems at school: this is a simple analog computer (see the description of Oughtred's circular slide rule in *Homo Problematis Solvendis*). In more formal terms, an analog computer is a device that uses continuously variable physical quantities, which can be electrical signals, the motion of mechanical components, or the pressure of a liquid, to solve problems. A very simple example would be an electrical circuit with two variable voltage inputs and a meter that measures the combined output of the circuit. If 1 and 2 V are input to this circuit, and the meter reads 3 V, then we have a simple analog computer that has calculated $1 + 2 = 3$. Of course, the circuit would have to be designed correctly to achieve this, but this is a simple enough task, and illustrates the point very nicely.

The Antikythera Mechanism was recovered in 1901 by divers, in a ship wreck off the coast of the Greek island of Antikythera. It is a mechanical, or clockwork, device consisting of at least 30 bronze gear wheels, the largest of which is 14 cm in diameter, with 223 teeth on the gear wheel. Thanks to meticulous scanning and imaging analysis, it has been possible to establish that the device—in a poor condition, thanks to its centuries on the sea bed (see Fig. 4.2)—was used to make complex astronomical predications. It is thought that these included calculating the positions of the Sun and Moon, and also predicting eclipses. More remarkable is that it did these things mechanically, over one thousand years before similar devices reappeared in Europe, in the fourteenth century.[2]

The period from which the device is thought to originate (approximately the first or second centuries BCE), and the location (roughly midway between the island of Crete, and the Southern tip of the Greek mainland), along with the artefacts

---

[2]The surviving pieces of the Antikythera Mechanism are held in the National Archaeological Museum, Athens. Image Credit: Tilemahos Efthimiadis. Original image in colour. https://creativecommons.org/licenses/by/2.0/deed.en.

Fig. 4.2 Antikythera Mechanism

found with the mechanism, make this undeniably an artefact of Hellenistic Greece.[3] Although long after the time of great philosophers such as Socrates or Plato, the mathematical knowledge that informs the device is known to have been possessed by Greek mathematicians and astronomers of the era. What is perplexing is that the device, a sophisticated mechanical computer, seems to appear rather suddenly in time, without any known antecedent, or immediately subsequent, devices. This seems to suggest not only that the knowledge required to make this device was very specialised, but also that the information it provided was very valuable, conferring (at least, potentially) some considerable advantages on the owner.

The Antikythera Mechanism allowed the user to compute a sophisticated range of astronomical information. It is tempting to assume that this information would serve a very practical purpose, like the navigation required for sea trade. However, seafaring of the period was typically conducted in sight of land. Whether hopping from island to island, or working around the coast of the Eastern Mediterranean, and North Africa, seafarers could make do with a basic knowledge of the night sky, the Sun, and other local geographic and meteorological features. If the Antikythera Mechanism was used for navigation, it is reasonable to expect that it would have been very common, in the same way that nearly everyone nowadays has a GPS feature on

---

[3] This is the period after the death of Alexander the Great (323 BCE) and the emergence of the Roman Empire (after the Battle of Actium in 31 BCE).

their smartphone. If it was very common, it is difficult to see how evidence of only one such device would emerge, and not until 2000 years after its creation.

We also know that the role and status of women changed markedly in the transition from the preceding *Classical* Greek Era, with women asserting greater rights, levels of education and economic freedoms. Greek women in this era emerged as philosophers in their own right and also held public offices. If the Antikythera Mechanism was not a common, navigational device, but a highly specialised religious or political device, we can begin to speculate not only that it may have been a rare and closely guarded secret, but also that it was developed by a woman responsible, for example, for calculating the dates of events of great religious, economic or social significance.

Can we narrow down the problem that the Antikythera Mechanism was intended to solve? From that, what can we learn about the creativity of this invention that seems to have been so far ahead of its time?

4.2.2 Why Was the Antikythera Mechanism Invented?

The obvious problem for which this innovation is a solution was the question of how to make a variety of astronomical calculations. However, these calculations served some higher purpose, and it must have been a purpose of considerable importance and value to warrant the investment in what seems to be a one-off device. These facts suggest that the purpose was more than navigation, especially in a region that could be well-served by relatively simple coastal navigation techniques. I suggested earlier that this special purpose may have been the prediction of significant religious, political, or social events that depended on accurate celestial computation. Even today, the setting of the date of Easter, in the Christian calendar, is determined by the timing of the first full Moon after the vernal equinox.[4] This was established in 325 CE, in an era without the benefits of modern calendars and other tools. Perhaps the Antikythera Mechanism was designed for a similar purpose? Whether we regard the problem as *how to make astronomical calculations*, or refine this to *how to predict celestial events*, we must now assess the creativity of this invention.

4.2.3 How Creative Was the Antikythera Mechanism?

Relevance and Effectiveness: As impressive, and apparently ahead of its time as the Antikythera Mechanism was, it has been speculated by modern scientists that it would not have been very accurate. I constantly remind myself, and the reader, that

---

[4]The vernal equinox is the date when the Sun crosses the equator, moving north. This traditionally marks the beginning of Spring in the northern hemisphere and is the date where night and day are equal in length.

we must put ourselves in the position of someone seeing these inventions when they were first introduced, and not allow hindsight to colour our opinions. However, if this invention is not accurate now, then it would have been inaccurate in 150 BCE. In other words, it is a little like the gadgets that we sometimes fall prey to in in-flight shopping magazines (or comic books, in my youth). They often look fantastic, and seem like a bargain, but fail to live up to their promise. *X-ray Specs* and *See-Behind Glasses* looked amazing in the comic book, but I am assuming they didn't actually work very well (if at all)! If relevance and effectiveness is a prerequisite quality of a creative solution—is the artefact fit for purpose?—then the Antikythera Mechanism is on shaky ground. Of course, this should not detract from its other qualities, but it damages its creativity. I feel that there is no choice but to give this innovation a rather mediocre 2.5 out of 4. It certainly calculates the positions of planets, and the dates of significant events, just not very well! Perhaps it ended up on the ocean floor because it was thrown there by an embarrassed civil servant?

Novelty: There are two elements to this criterion, and together, both form the second prerequisite for product creativity. First, we ask if the invention in question helps to define the problem at hand. In the sense that the Antikythera Mechanism shows how other solutions—e.g. old-fashioned star-gazing and manual calculations—could be improved, it must score highly. It goes without saying that a machine capable of performing these calculations more or less automatically would be enormously beneficial. Indeed, this is the basis of many modern devices like computers. They don't necessarily do things that we cannot do, they just do the things we can do very, very quickly, and with greater accuracy. On the downside, however, it is hard to claim that this innovation highlights the weaknesses of other solutions when it doesn't really work very well. The novelty of the Antikythera Mechanism is therefore somewhat *latent*. It is waiting to be realised.

The second aspect of novelty that is important is the extent to which the innovation sheds new light on the problem. Once again, the picture for this artefact is a little mixed. In some respects, it is strong, doing well as a radically new approach to an existing problem. In other respects, it is a little weak, for example in terms of how an observer might have seen new and different ways of using the artefact. In this latter case, I find it difficult to imagine a person at the time thinking "you know, I could also use this to calculate my taxes", or "this system of gears could also be used to fashion a clock". If that seems somewhat unfair of me, then consider the fact that the Antikythera Mechanism (and devices like it) seemed to disappear from the historical record for over a thousand years. Perhaps the problem is that it was *too* novel. It was so far ahead of its time that people couldn't imagine other uses for it. I have given it a score of 3 out of 4 for novelty. Not bad, but it misses maximum points probably through no fault of its own but more due to the inability of our forebears to exploit it to a greater degree.

Elegance: Although we cannot say for sure, because of the heavy corrosion to the device, and the lack of any other specimen, the Antikythera Mechanism was *probably* a very impressive looking machine. However, this also reminds us that elegance is not the same as simple aesthetic appeal. This criterion acknowledges that a good solution usually *looks like* a good solution (but looks can be deceiving)!

As impressive and convincing as it may have looked, and as apparently complete, graceful and harmonious as the design may have appeared, it seems that this was not enough to ensure good performance. I give it a maximum score of 4 out of 4 for elegance, but with some reservation. What we typically find is that high elegance is an indicator of high effectiveness. In the case of the Antikythera Mechanism, this is not the case. I wonder whether the maker of this device knew that the performance was not quite up to scratch and was trying to draw attention away from this by making a *beautiful* artefact?

Genesis: A consistent theme for the Antikythera Mechanism has emerged. In some ways, it is very impressive, but in others, it may have been let down by poor execution. The same applies to genesis. In many respects, this invention does change how the problem was understood. Not only were the movements of planets and stars predictable, but this predictability could be *programmed* into a machine. Its status as the first analog computer is unchallenged, and yet, without the necessary accuracy, it remains an excellent *concept demonstrator*, rather than a viable solution. If the engineers of the day had possessed the supporting technologies to refine this innovation, then we might have seen mechanical clocks, control mechanisms, and other mechanical devices hundreds of years earlier than we did. Unfortunately, the qualities that help to define genesis—e.g. vision, seminality, and germinality—require something that is a more convincing solution to the underlying problem. 3 out of 4 means that it is not quite the paradigm breaker that it could have been. Once again, this may be simply because it was too sophisticated for its day, and too far in advance of its time.

Total: The Antikythera Mechanism score 12.5 out of 16 for creativity. Like its Classical Era companion, the Construction Crane (see *Homo Problematis Solvendis*), it thus sits at the junction of *high* and *very high* creativity. The culprit, in this particular case, is effectiveness. Not, as is often the case, due to poor execution, but actually despite good execution! This is almost certainly because the idea exceeded the capacity of the people of the time to successfully execute it. Rather than a criticism, that's really a compliment. Whoever invented this device—some unknown *Femina Problematis Solvendis*—was a true genius, conceiving and prototyping an analog computer centuries ahead of its time!

4.3 The Bain-Marie (0–300 CE)

At its essence, art is an alchemical process. Alchemy is a process of transformation—Julia Cameron, American Teacher (1948–)

The final invention in the Classical Period is the first that we can attribute unequivocally to a woman. This does not entirely resolve the question of *why her*, but it does remove some of the guess work! Of course, we shouldn't ignore the fact that asking *why her* also begs the question of *why him*, in the case of inventions by men (or indeed, why *not* her?). We know that the answer to this, at least for some periods of human history, was that society didn't *permit* her. However, even in the twenty-first century, some obstacles for women remain—we see these, for example, in lower participation rates in STEM subjects by young women—that continue to impact on creativity, innovation, invention, and problem solving.

Nevertheless, in the case of the bain-marie, a simple energy-handling system, we can ask why Mary the Jewess—often considered the first true Western alchemist[5]—devised this invention, and what problem she was addressing by doing so.

We know of Mary the Jewess principally through the writings of Zosimos of Panopolis, an Egyptian alchemist who lived in the third or fourth century of the Common Era. This also explains the uncertainty in the date of this invention. In his books, Zosimos describes various alchemical instruments and experiments attributed to Mary, whom he holds in high esteem, referring to her as "one of the sages". We will focus on what is perhaps her most famous innovation: a simple, but effective, mechanism for controlling the temperature of a heat source in an era lacking the sophisticated control systems that we take for granted in the modern world.

4.3.1 What Was Invented?

The bain-marie, or bain de Marie (from the French for *Mary's bath*, see Fig. 4.3) is simply a *double boiler*. In practice, this means that an outer container is filled with water, and an inner container is partly immersed in the water. The outer container is heated, causing the water it contains to warm. The heat in this water is then transferred to the inner container, heating whatever is contained in the inner container. If that seems rather complicated or unnecessary, keep in mind that the water in the outer container can only be heated to boiling point (100 °C at sea level). Once the temperature reaches 100 °C, the liquid water changes to steam. Instead of getting hotter, the water uses the energy it is absorbing to change state.[6] This means that the inner container, sitting in this boiling water, can only be heated up to, but not beyond, the same temperature. The advantage of the bain-marie is that it allows substances to be heated, or maintained at 100 °C, in a more controlled and precise manner. For cooking, this is very handy—the contents of the inner container cannot easily be

---

[5] As many readers may know, alchemists were, in effect, early chemists. Alchemists practised alchemy, and that word derives from a Greek term meaning "art of transmuting metals". The alchemists were renowned for trying to change—transmute—base metals into gold. While they do not appear to have succeeded, they did develop many tools and techniques important in the science of chemistry.

[6] If you did science at high school, you may well have conducted an experiment to observe this. You heat water in a flask and monitor its temperature. The more you heat it, the more the temperature rises. However, at 100 °C, the temperature stops rising, even though you are continuing to heat the water. The heat you are adding is now not being used to raise the water's temperature, but instead is freeing the water molecules from their liquid state and giving them enough energy to allow them to boil away. Provided you don't boil away all of the water in the flask, its temperature will remain steady, giving you a stable, controllable heat source.

Fig. 4.3 Example of a
Bain-Marie (1528)

burned or scorched—so a similar arrangement is used to reheat milk for babies, or to melt chocolate, even today.[7]

Mary, of course, was not using her innovation for cooking, but for alchemy. This ancient form of natural philosophy,[8] which originated in Greco-Roman Egypt at some time in the centuries following the birth of Christ, was really an early form of systematic *science*. Principally concerned with what we would now think of as a combination of chemistry and medicine, alchemy sought, among other things, to transform *base metals* (common and inexpensive metals such as lead) into *noble metals* (e.g. silver and gold). Alchemy, and alchemists, also searched for an elusive *elixir of immortality* (also referred to as the elixir of life), cures for various ailments, and even the search for a universal solvent (or *alkahest*) capable of dissolving all other substances. In the European tradition of alchemy, which included famous scientists such as Isaac Newton (1643–1727) as a practitioner, some of these alchemical objectives merged into the famous *Philosopher's Stone*[9]—a substance that could turn base metal into gold and was also the elixir of life.

Although alchemy traditionally had a strong *esoteric* element to it—that is, a spiritual or mystical component—it also had a strong *exoteric* (i.e. objective and

---

[7] Image Credit: From *Coelum philosophorum, seu De secretis naturae liber/ Philippo Ulstadio Patricio nierebergensi authore* by Philippus Ulstadius. Argentorati: Arte et impensa Joannis Grienynger, 1528. Science History Institute, Philadelphia. Public Domain.

[8] The precursor to natural science, i.e. the study of nature and the physical world.

[9] Fans of Harry Potter will be familiar with this.

observable), or what we might call *scientific*, tradition. Regardless of the underlying objectives and motivations, it is not hard to see that alchemists needed tools, methods, and techniques to support their work. Indeed, modern chemistry continues to make use of laboratory techniques, terminology, and experimental procedure first developed by alchemists. For example, alchemy introduced the tools and method of distillation[10] to western Europe. Alchemists of the Middle Ages also developed sulphuric and hydrochloric acids, and gunpowder was a Chinese alchemical invention.

The bain-marie is an excellent example of an instrument developed for alchemical purposes, and which has found more diverse uses. Aside from culinary applications, the device may be used to heat wood glue. This glue must be maintained in a liquid state, but its adhesive properties can be damaged if it is heated too strongly. Very simply, any process that requires precise and controlled heat, and where modern control techniques are either unavailable or unfeasible (e.g. due to cost) can benefit from the bain-marie. Even boiling an egg represents a crude form of this double-boiler system!

4.3.2   Why Was the Bain-Marie Invented?

Our ancient ancestors must have understood that water boils at a fixed temperature (allowing for minor changes associated with altitude). They must also have understood that boiling represents a change of state, and that the boiling water in a heated pot will eventually disappear. It is equally certain that they knew that wood—their primary source of heat—burned at a much higher temperature. One of the difficulties of wood as a source of heat is not only that it *begins* combusting at about 300 °C, but that it burns over a larger range of temperature. A wood fire may therefore range from 300 up to about 600 °C. This is not only too hot for many alchemical applications, but also hard to control with any precision. The problem that Mary the Jewess was seeking to solve was *how to control heat*.

4.3.3   How Creative Was the Bain-Marie?

Relevance and Effectiveness: The bain-marie worked well. It was a simple, feasible idea that did not require any particular extension of the available technology. It did what it was designed to do, and did it well, perhaps because the underlying principle was very simple. I have no hesitation in giving it a 4 out of 4 in this category.

Novelty: The bain-marie emphasises the inherent uncontrollability of wood fires, as well as the fact that wood burns at a temperature too hot for many applications. However, it didn't really show how existing artefacts/solutions could be improved,

---

[10]Distillation involves purifying a liquid through a process of heating and cooling and uses specially designed apparatus to achieve the necessary stages.

because there really weren't any alternatives beyond a pile of burning wood. The device itself doesn't really extend the known, but is a radically new approach, and also quickly prompts observers to think of other uses. What else needs to be gently heated? Food, milk, glue? The bain-marie is worthy of a strong score of 3.5 out of 4 for novelty.

Elegance: Surprisingly, I come out a little tougher on this invention when judging elegance. This is surprising, in part, because we normally expect to see a high effectiveness score accompanied by high elegance. Although we cannot know for sure, we may speculate that the execution of the bain-marie left a little to be desired. If engineers struggled to make a reliable steam pump in the 1600s (see *Homo Problematis Solvendis*), then it is conceivable that Mary and other technicians of this much earlier era would have had only rudimentary metal-working skills. The device depicted in Fig. 4.3 requires some fairly sophisticated metal-working and/or glass-making skills. While not *beyond* the technology of the day, it may have stretched the technical abilities of people at the time. They could make a bain-marie, but I suspect it may have leaked, or had unreliable seals and other key components. Fortunately, it probably worked quite well because the concept was fairly simple, *in spite of* the execution, and not so much *because of* it. Therefore, I will give it a score of 2.5 out of 4.

Genesis: The bain-marie is important not only for what it does, but also because it has a strong element of shifting thinking about the nature of the underlying problem. Prior to this innovation, our ancestors had only one avenue for controlling heat, both in general, but especially for more delicate processes such as cooking or alchemy. Heat, usually obtained from the combustion of wood, or possibly oil, could be varied only by moving the heat source closer to, or further away, from the object to be heated. That gave some degree of control from the maximum temperature of the burning fuel, down to lower temperatures. The weakness of this approach was that heat is transferred to the object by radiation, so that the degree of control is haphazard. Without an accurate thermometer, the heat applied to the object of interest could only be estimated. For temperatures higher than the combusting fuel, some control could be obtained using, for example bellows. A blacksmith could thus heat metal with some degree of accuracy. The problem of controlling heat, especially at *moderate* temperatures (e.g. between room temperature, and the temperature at which wood begins to combust), was particularly severe. The paradigm shift of the bain-marie was not only to control heat via an intermediate material (water, in the case of the bain-marie), but also to transfer the heat by convection and conduction, instead of radiation. Perhaps without even realising it, Mary the Jewess opened up this expanded understanding of the nature of heat, in addition to solving the immediate problem of providing precise control of that heat. Of course, the bain-marie wasn't perfect, and had one setting—100 °C! However, as with many innovations, the conceptual breakthrough had been made. I give the bain-marie 3 out of 4 in this category.

Total: With a total score of 13 out of 16, Mary the Jewess's invention sits just inside the *very high* range for creativity. The only major weakness appears to be in execution of the solution. Rather like the steam pump of a later era—a groundbreaking idea that was somewhat ahead of its time—the bain-marie was held back slightly by the fact that the technology of the day couldn't fully rise to the demands of this device. This might have had a more serious impact on the invention, but it appears that the device was so simple that any problems of execution did not overly affect its performance.

Chapter 5
The Dark Ages (476–1453 CE): Medieval Creativity

The next in our sequence of time periods is the span of approximately 1000 years known (in Western culture) as the Dark Ages. This epoch spanned the fifth to fifteenth centuries of the Common Era, beginning approximately with the sacking of Rome—and the fall of the Western Roman Empire—in 476 CE, and ending with the fall of the Eastern Roman Empire[1] centred on Constantinople (modern-day Istanbul), in 1453 CE.

The very name of this period, the Dark Ages, suggests that we might not find much in the way of human creativity and innovation. However, while some of the advances, organisation, and sophistication of the Western world evaporated with the sacking of Rome, neither the world in general, nor even Europe, stood still. Many important inventions were created across this time period, not just in Europe, but across the globe. The Chinese, for example, invented the horse collar in around 500 CE, allowing those animals to pull much larger loads. It was also in this period that the Chinese invented gunpowder, woodblock printing, tower clocks, and cannons. The Arab World, in the same epoch, invented optical lenses, the sextant, the crankshaft, and an early form of mechanical robot.

We will, however, stay in Europe and begin with an invention that has been critical to our ability to communicate, ever since the invention of writing—the *Quill Pen*. The second innovation that we consider in this epoch is then a direct offshoot of the first and is a constructed language called *Lingua Ignota*. Our final invention for the period of the Dark Ages is the invention of the *Verge Escapement*. Our earliest ancestors understood the utility of rotational motion, from the simple Wheel through to the Construction Crane. However, one trick to making better use of rotational motion was the question of how to *control* that rotation. This mechanical device achieved that control, allowing rotation to be used for functions such as timekeeping.

Although in the Dark Ages we begin to see the clear, recorded emergence of female inventors, the era also highlights some of the inequalities that have impeded *Femina Problematis Solvendis*. Once societies began to divide roles along gender lines, whether driven by the advent of agriculture, or by something else, we entered

---

[1] Also known as the Byzantine Empire, or Byzantium.

© Springer Nature Singapore Pte Ltd. 2020
D. H. Cropley, *Femina Problematis Solvendis—Problem solving Woman*,
https://doi.org/10.1007/978-981-15-3967-1_5

a phase where women were either explicitly or implicitly denied the opportunity to invent. You can't do that because we (men) won't allow you to—explicit denial—or, you can't do that because you lack access, for example, to education—implicit denial. These two barriers would exist for centuries, only really beginning to be dismantled with small changes in the nineteenth century, before major improvements to gender inequality began seriously in the mid-twentieth century.

5.1   The Quill Pen (c580 CE)

The pen is mightier than the sword—Edward Bulwer-Lytton,[2] English Author (1803–1873)

The quill pen is thought to have been invented late in the sixth century CE, and probably replaced the reed pen. Although paper was invented in China in 105 CE (see *Homo Problematis Solvendis* for a discussion), reaching Europe in the early 1100s, we cannot directly link the two inventions, as the dates suggest. The reason, of course, is that other materials (e.g. parchment, or silk) were widely used as writing *materials* for centuries before the advent of this writing implement.

What we do not know is *who* exactly invented the quill pen. If we apply our *why not a woman* rule, then it is interesting to speculate about the circumstances that might have led to one of our ancestral mothers developing this writing implement. It has been suggested that the quill pen first appeared in what is modern-day Spain in the period known (at least in European history) as the Dark Ages. The date of the quill pen predates the Moorish invasion[3] of Spain in 711 CE, meaning that the culture at the time was a Romanised (and Christian) Celtic culture, much like what existed in Britannia (modern Britain) and Gaul (modern France). The critical thing this tells us is that writing was a rare occupation, and probably limited to small numbers of educated monks, and, for want of a better term, bureaucrats. Unfortunately, in the era in question, this meant men. So, if men were doing the writing, what circumstances led a woman to the idea of a quill pen? I can imagine a scribe, working late at home, trying to finish a document by the light of a candle. In his haste, he breaks his reed pen, or perhaps it has become soft and prone to leaking ink onto his precious document. Perhaps he flung the reed aside, cursing that "if only there were a better pen that could hold the ink, and make a better mark on my parchment!". I then see his wife, perhaps having just plucked a goose for their next meal, saying "Well what about this feather? It's sort of like the reed pen, and I've noticed that it is hollow. Maybe you could make a pen from this?" If he was anything like me, he probably dismissed the idea, and sat cursing his reed pen. In the meantime, his enterprising

---

[2]Edward Bulwer-Lytton, coincidentally, was the romantic partner of Henrietta Vansittart, whom we shall encounter in a later chapter. Vansittart invented the Lowe-Vansittart propeller in 1869 and conducted an affair with Bulwer-Lytton over a number of years.

[3]The Muslim Moors invaded Spain from Africa, crossing the Strait of Gibraltar, with an army under the command of Tariq ibn-Ziyad.

(and calmer) wife went to the kitchen, took a knife and fashioned the first quill pen. A little experimentation, and like many inventions, a new solution was born.

Both on its own, and together with various forms of writing material, the quill pen forms an information-handling system. I have previously argued that paper could also be regarded as a material-handling system. The same argument could be made for the quill pen, but I prefer the idea of information-handling as a better representation of the true value of this, and other, writing implements.

5.1.1 What Was Invented?

Quill pens are made from the moulted (i.e. the shed) feathers of large birds. However, not all feathers are the same, and the preferred type for quill pens are the primary wing feathers of birds such as geese. Quills are also made from swan feathers, as well as those of crows, eagles, owls, hawks, and turkeys. The choice of species is typically determined by the size of the quill, with crow feathers, for example, favoured for fine work. The wing feathers are an example of the more general type known as *vaned* feathers, in contrast to *down* feathers, and have the vital hollow, tubular quill that is critical for the function of the pen (see Fig. 5.1).[4]

Fig. 5.1 Seventeenth century painting of a scholar cutting his quill pen

---

[4]Image Credit: Gerrit Dou, Geleerde die zijn pen snijdt (Scholar cutting his pen), 1630–1635, The Leiden Collection, Public Domain.

Quill tips, i.e. the end doing the actual writing, could potentially be cut in a variety of ways to suit particular types of writing or drawing; however, a square-cut, and rigid, tip was common up to the sixteenth century. After this time, and as writing became more widespread, *copperplate* script became a preferred style of handwriting characters. This led to changes in the shape and rigidity of the quill tip, with a more pointed and flexible tip becoming the favoured style.

The quill pen works by holding a small reservoir of ink inside the hollow shaft of the quill (feather). The tip, aside from being shaped either square or with a finer point, also has a slit cut into it running along the length of the shaft for a short distance. When the writer presses the quill tip on the writing surface, the shaped tip splays apart very slightly, thanks to the slit. This, in effect, opens up the tip very slightly, and through the magic of *capillary action*,[5] allows some ink to flow from the reservoir into the tip, and onto the writing surface. If you are old enough, or perhaps *old-fashioned* enough, you may have used a more modern version of the ink pen at school. Nowadays, the ink is usually held in a small, replaceable cartridge, and the tip is metal, but the capillary action is exactly the same as the quill pens our forebears used hundreds of years ago.

5.1.2   Why Was the Quill Pen Invented?

The underpinning problem that the quill pen was developed to solve seems to be something like a question of how to communicate. This is not a bad attempt at stating the problem; however, it has some minor flaws. Our best practice at problem definition has been to use the form *how to verb noun*. The statement *how to communicate* lacks the noun. This may not seem like a major issue; however, the noun plays a very important role in the problem statement. It is the noun that focuses, or constrains, our problem. Without any noun, the problem space becomes too large. What do I have to communicate? Conversely, if the noun is too specific—how to communicate verbal agreement, for example—the problem space may be too narrow. An infinite problem space can make it hard to focus on any solution, just as a very narrow problem space may make it hard to focus on anything other than one solution. So, we need a *Goldilocks* noun: not too vague, not too narrow, but just right! My usual advice is that if we have a noun, and if that noun is fairly open-ended, or non-specific, then we are probably in a good spot. My go-to example if to think about the problem a screwdriver is solving. *How to screw*, a verb and no noun, is not only open to double meanings (!), but begs the question "screw what?" *How to screw-in a screw* now has a noun, but both the verb and the noun are too narrow and specific, limiting possible solutions. If we are trying to find a range of creative alternatives to the screwdriver,

---

[5]Capillary action is the ability of a liquid to flow in narrow spaces, like the inside of a feather, without the aid of any external forces (e.g. gravity). It is the mechanism by which oil is drawn up into the wick of a lamp (hence why the term *wicking* is also used), or water can be drawn up into a strip of paper.

then the best problem statement is probably something like *how to apply torque*.[6] Back to the quill pen, where I think a good statement of the problem is *how to communicate in writing*. If we were also to add a *how well* (a constraint), then it might be supplemented with *quickly and clearly*, or something similar.

5.1.3 How Creative Was the Quill Pen?

Relevance and Effectiveness: It is clear that the quill pen did not exceed the technological capabilities of its time. This is an interesting case, because it is not entirely a human invention. I am not suggesting that the quill pen falls into the category of *discovery*, but it blurs the lines between invention and discovery somewhat. So, while it was correct and feasible, it might be argued that it was not a perfect solution to the problem of writing and written communication. In particular, it must lose a little for the fact that it required some skill and experience. A person could not simply pick up a quill pen, having never used one before, and start writing clear, flowing text. Not only did it take practice, which falls foul of our constraint *quickly*, but the risk of spilling ink, smudging it, or making a mistake limits the constraint *clearly*. In the hands of an experienced scribe, there is no doubt that it was highly, but not perfectly, fit for purpose. I therefore give the quill pen 3.5 out of 4 in this category.

Novelty: the quill pen is reasonably strong across all elements of novelty. It draws some attention to the shortcomings of other writing implements, whether reed pen, lump of charcoal or a stylus scratching characters into a wax tablet. The quill pen, like a reed, continued to shift our ancestors' concept of the problem of writing. Rather than scratching marks into a surface, the idea of applying a substance to a surface was taken a step further. The quill pen was not merely a kind of paint brush, transferring pigments from one place to another, but a more self-contained unit, able to create very fine, controlled lines, with a much simpler and less viscous ink. The quill pen contained the seeds of later improvements: what if the pen could carry all of its ink with it? What if it could apply that ink to all sorts of surfaces? The quill pen is not merely incremental in nature but begins to hint at more radical changes. I give the quill pen a score of 3 for novelty.

Elegance: The quill pen is a curious invention in this category, perhaps again driven by the fact that it is not entirely a human invention. Our ancestral *Femina Problematis Solvendis* did not create the quill pen from nothing, but cleverly *adapted* something that already existed. That is no less impressive. It still requires the ability to formulate the underlying problem, and to abstract the desired functionality, before seeing that functionality in another object. In this category, however, we cannot entirely attribute the elegance to the work of a human. The long, successful history of the quill pen suggests that it was, broadly speaking, well-executed. However, some of the characteristics of elegance, e.g. completeness or harmoniousness, may be as

---

[6]Now it is easy to see that the solution could be a screw-driver, but it could also be your fingers, a coin, a credit-card, a knife, a piece of string, and so on.

much due to *Mother Nature*, as they are to our inventive, ancestral mother. On the plus side, however, is the fact that part of our creative ability as humans is the ability to recognise good solutions. That means not only recognising good solutions in what we invent, but also recognising the potential of things in the natural world to be good solutions. This calls to mind the field of biomimetics,[7] and the ability to adapt solutions from nature to our human problems. For elegance, I give the quill pen 3.5 out of 4.

Genesis: The quill pen is quite strong in this category. It contains elements that could, and later would, serve as the foundation of further developments. It has some degree of transferability, for example in the application of capillary action. It also must, at some point, have caused people to ask if there were easier ways to achieve the same outcome. In general terms, did it break the writing paradigm? Although it has many strong qualities in this category, the answer to this question is an equivocal *sort of*, or *not quite*. It was not merely incremental, but it was not strongly radical, and for this reason scores a good, but not outstanding, 3 out of 4 for genesis.

Total: Sometimes I begin these case studies full of hope for an invention, expecting a really high score for creativity, but ending up a little surprised, and even disappointed. That is, of course, the point of these assessments. All our inventions are, in the grand scheme of things, creative. However, when we peer deeper into their qualities, we find nuances and details that help us to understand a little better the nature of creativity and how it is manifest. The quill pen, as an example of outstanding creativity is, in my view, not quite there. With an overall score of 13, it sits just inside the very high range. And yet, it is just a little deficient in each category, in ways that just hold it back. If it had been easier to use, and more accessible to anyone wishing to write (in the way the modern ballpoint pen is), or if it had sparked improvements and refinements sooner, then it might be one of our most creative inventions. Instead, it is clear that it did a job that was good enough for our forebears to be generally satisfied with it, and for centuries. That is quite a compliment to the invention, and its inventor, but it leaves me wondering why it didn't develop further sooner. At 13 out of 16 for creativity, it is something of a curiosity in our catalogue of inventions.

5.2   Lingua Ignota (c1200)

There was no light nonsense about Miss Blimber ... She was dry and sandy with working in the graves of deceased languages. None of your live languages for Miss Blimber. They must be dead—stone dead—and then Miss Blimber dug them up like a Ghoul [Dombey and Son, 1848]—Charles Dickens, English Writer (1812–1870)

Hildegard of Bingen, a member of the Order of Saint Benedict, was born in what is now the Rhineland-Palatinate region of modern Germany, in approximately 1098.

---

[7]The field of biomimetics, or biomimicry, involves imitating structures, systems and methods found in nature for the purpose of solving human problems. See the book *The Shark's Paintbrush* (2013), by Jay Harman.

In an era, and a region where we might be tempted to assume that women were universally subordinate to men, disenfranchised and generally lacking in economic, social or political power, Hildegard of Bingen stands out, like some of her contemporaries (e.g. Eleanor of Aquitaine[8]), as a powerful and influential figure in the High Middle Ages. Hildegard was a polymath, in the same sense that we use the term to describe people like Leonardo da Vinci. She was renowned as a theologian, a writer on scientific and medical topics, a composer of sacred music, and also as a prolific correspondent with Popes, Emperors and other contemporaries in the Church. Hildegard was also known for the sacred visions that she experienced. These began as a young child and continued throughout her life. Although hesitant for many years to share these with others, she recorded many of her experiences after Pope Eugenius (1145–1153) heard of the phenomenon and gave his approval for them to be recorded and shared. Among all these considerable achievements, and her establishment of two monasteries, Hildegard also found time to invent a new language: *Lingua Ignota*.

5.2.1 What Was Invented?

In the book *Homo Problematis Solvendis*, I described the invention of Cuneiform. Not a language, but an *alphabet* that was used to write down, for example, Akkadian: the language spoken in Mesopotamia in the period around the middle of the third millennium BCE. Most languages have arisen naturally, and over long periods, evolving and changing along the way. English is a good example, tracing its origins back through a complex path tied to Frisian, Dutch, and German, and strongly influenced by the movement and mixing of people over many centuries. We can see how English is related to these other languages, and get a sense of their common ancestry, if we consider words like "day" in English ("dei" in Frisian, "dag" in Dutch and "Tag" in German).

Lingua Ignota[9] in contrast, is what is known as a *constructed* language—one that is deliberately invented—and used both an invented alphabet (23 so-called *litterae ignotae*, or unknown letters, see Fig. 5.2), and invented words, but with existing Latin grammar, to form a new language. A language, therefore, is a system of spoken communication, which through the means of an alphabet, can be written as well.[10]

The only existing text in Lingua Ignota is the following passage. It is written, in this case, in the Latin alphabet (obviously, the alphabet in which I am writing this text) and contains only five words that are unique to Lingua Ignota (shown in bold):

---

[8]Eleanor was Queen Consort of France from 1137 to 1152, after marrying Louis VII. When her marriage to Louis was annulled, she then married the Duke of Normandy, who became Henry II of England, from 1154 to 1189.

[9]From the Latin for *unknown language*.

[10]From Lingua Ignota per simplicem hominem Hildegardem prolata, by Hildegard of Bingen, circa 1200. Image Credit: Public Domain.

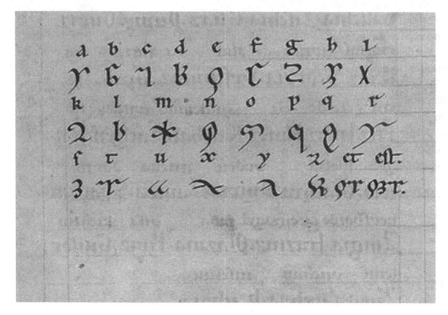

Fig. 5.2 Litterae Ignotae

*O **orzchis** Ecclesia, armis divinis praecincta, et hyacinto ornata,
tu es **caldemia** stigmatum **loifolum** et urbs scienciarum.
O, o tu es etiam **crizanta** in alto sono, et es **chorzta** gemma.*

Only one of these invented words is in the known glossary (Loifol = people). The other four uniquely Lingua Ignota words are not known to us.

5.2.2   Why Was Lingua Ignota Invented?

The question of *why* Hildegard of Bingen invented Lingua Ignota remains unclear. Past theories believed the language was invented with the intention of forming a universal language. In other words, it was thought that Hildegard sought to create a single language through which all people could communicate. Some readers may be familiar with, perhaps even speak, a similar language: Esperanto. This was created in the late nineteenth century with a similar goal. Although Esperanto claims as many as two million speakers in the world today, my personal impression is that it has struggled to make inroads as a universal second language. At school in the 1970s, I heard about it. Nowadays, however, it never seems to be mentioned.

Current theory, however, believes that Lingua Ignota was intended as a secret language. We know that Hildegard of Bingen was an avid communicator with many influential individuals in positions of power. Perhaps Hildegard was seeking a way of communicating with these people in code? The history of Europe in the High Middle Ages, as with later periods, is full of political and theological intrigue. Kings and Queens ruled often divided realms, with rivals seeking to stake their own claims to power. Countries regularly fought over disputed lands. On top of all this, the Church, in this era, exerted an enormous influence of its own throughout Europe. Was Hildegard of Bingen an influential powerbroker behind the scenes? Did her position in the Church, and possibly her gender, necessitate a degree of secrecy and discretion that a coded language could provide?

The uncertainty regarding *why* she invented this language has important consequences for our assessment of creativity. If Hildegard invented it in the hope of creating a universal language that would usher in peace and cooperation, we will judge its effectiveness very differently compared with the creation of a secret language, only intended to be comprehensible to a select few. If her goal was the former, then the problem was how to *facilitate* written communication. If the latter was her objective, then the problem she sought to solve was almost the opposite: how to *restrict* written communication! We will go with the latter, since Lingua Ignota did not become widely used, and we know very little even about some of the words she coined.

5.2.3   How Creative Was Lingua Ignota?

Relevance and Effectiveness: Although it has been my *modus operandi* throughout this, and the preceding volume (*Homo Problematis Solvendis*), I am going to depart briefly from sticking to one clear, declared problem. How can we judge the creativity of a solution to a problem if we are uncertain about the problem? However, in the case of Lingua Ignota, I want to highlight the differences in creativity that result from a subtle change in the nature of the problem. If Hildegard's intention was to create a widespread, universal language, then Lingua Ignota was not very effective. Not only did it require a good understanding of Latin grammar—something that only a small number of people possessed at the time—but it required learning a new alphabet, and new words composed in this alphabet. In an era when only priests and the elite of society were likely to have the necessary education, this would have doomed Lingua Ignota from the start. As a universal language, it could only receive a weak score of, perhaps, 2 out of 4.

On the other hand, if Hildegard of Bingen was attempting to create a secret language that only other political and theological leaders could understand, then what better way to start than to base it on a foundation that was already familiar to her audience, i.e. Latin? Base it on Latin grammar, so that the intended users can quickly grasp the structure, but tweak that Latin with a modified alphabet, and some invented words, and suddenly you have a code that is both easy for intended users to pick up, and hard for unintended users to crack. In this case, 4 out of 4 seems like a fair score. We know that it was hard to decipher because we are unable even today to interpret some of the invented words! It was so good at encoding and limiting written communication that it died out completely once Hildegard and her confederates stopped using it. As a solution to the problem of a secret language, we will stick with a score of 4 out of 4.

Novelty: With the greatest of respect to Hildegard of Bingen, I have given Lingua Ignota a modest 2.5 out of 4 in the category of novelty. Her invention was, as we have seen, effective, but can we really say that it was highly original? Aside from new letters and some invented words—both rather incremental changes—we could argue that it differed very little from its Latin relative. It was, in essence, a rather elementary substitution code (or cipher), of a type that was easily broken even in the days of Napoleon.[11] Of course, it was developed many centuries before our modern understanding of codes and ciphers, so that the score of 2.5 reflects modest, but not outstanding, novelty.

Elegance: In contrast to its novelty, Lingua Ignota returns to a strong position when we consider its elegance. Even only looking at the characters in Fig. 5.2 we can see the care that was taken to create the *Litterae Ignotae*. These were not random scribbles, but gracefully executed symbols. A text written using these letters would have been immediately recognisable as a piece of written communication, even if it could not be interpreted. We are all familiar with seeing text written in, for example, the Greek alphabet, or in Cyrillic. Take γεια σας or Привет—both are the word "Hello" written in Greek and in Russian. Even if we do not speak these languages, most of us would probably guess that they were words written in an unfamiliar language, and not simply assume that they were meaningless squiggles. Lingua Ignota *looked like* a language, in the form of its letters, and in the structure of its grammar. Good solutions usually look like good solutions, and Lingua Ignota is no different in this respect from a well-constructed bridge or a sleek and aerodynamic racing car and deserves a high score of 4 out of 4 for elegance.

Genesis: One of the difficulties of a constructed language is that it may lack the organic and natural evolution of grammar, words, letters and even style that natural languages enjoy. Normal languages—English, Latin, Arabic, Chinese, to name only a few—did not spring into being, fully formed and mature, in the blink of an eye. All of the languages that humans speak have developed over hundreds or even thousands of years. Some have changed much more than others. The English of Geoffrey Chaucer's Canterbury Tales (1392), for example, is recognisable, but very different from modern English. Chinese, on the other hand, has been relatively stable for a much longer period of time. The problem with Lingua Ignota is that, being artificially and rapidly constructed, it must borrow heavily from some source. Almost by definition this seems to place severe limits on just how ground-breaking it can be as a language. Lingua Ignota did not change the underpinning grammatical paradigm, because it imported Latin grammar. It made minor, but certainly not radical, changes to the alphabet, and to some words. It is hard, therefore, to make a case that it changed

---

[11]For an excellent account of code breaking during the Napoleonic era, read Mark Urban's book *The Man Who Broke Napoleon's Code*. This is the story of Englishman George Scovell, who was in charge of the Duke of Wellington's efforts to break and read French military and diplomatic ciphers during the Peninsular War in the years preceding the Battle of Waterloo.

how the problem of written communication was understood. In retrospect, it teaches us something about codes and ciphers, but that is about all. I am therefore unable to score this invention at more than 2 out of 4 for genesis.

Total: I'll break my rule again and give both scores: if the intention was to create a universal language, then Lingua Ignota probably scores, at best, 8.5 or 9 out of 16. It was ineffective in this respect, lacked novelty, was inelegant (if it was no easier to learn than Latin) and incremental at best. However, our real focus is as a means of highly restricted communication: we could almost consider it a cipher rather than a language in the strict sense.

As a secret language, it finishes with a score of 12.5 out of 16. As we frequently see, effectiveness and elegance go hand in hand. An elegant solution is an effective solution, and vice versa. Where Lingua Ignota loses some points is on novelty and genesis. It is an incremental change to Latin: a fresh set of characters, and some new words, rather like a makeover! This is not to diminish a rather good overall score. 12.5 places this innovation in the space between high and very high creativity. Greater novelty, perhaps in the form of a unique grammar, would have bumped this up, as would a more sophisticated system of substituting new symbols for existing characters. Remember, as a solution to the problem of *restricting* communication, this would have been an easy code to break. Greater novelty and genesis applied to *Lingue Ignota as a cipher* is where the greatest potential for higher creativity is lacking.

5.3 The Verge Escapement (c1300)

What is better than wisedoom? Womman. And what is better than a good woman? Nothyng—
Geoffrey Chaucer, English Writer (c. 1340–1400)

We end this epoch with a simple, but important, energy-handling system. In *Homo Problematis Solvendis*, I suggested that the clock/pendulum was an information-handling system. Here, however, I think it is right to look at the verge escapement as the energy-handling system that then makes the subsequent information handling system possible.

It is not known who invented the verge escapement, but this mechanical device was critical to the development of modern, reliable timekeeping systems. Although mechanical clocks would undergo further refinement over the next 350 years, by 1656 they had reached a level of sophistication that would see the pendulum clock remain the most accurate form of timekeeping until the 1920s (see *Homo Problematis Solvendis* for a discussion).

Before the invention of this device, and mechanical clocks, how did our ancestors keep time? For most people, their daily lives were governed by simple natural cues: sunrise and sunset, for example. However, as medieval societies gradually transformed from primarily agricultural, to an increasingly diverse and integrated set of economic activities, trade and commerce, the coordination of the different parts of this system became more important. We see the importance of accurate timekeeping very clearly, before the end of the Renaissance, in sea trade and navigation.

5.3.1   What Was Invented?

The verge escapement, also sometimes referred to as the crown wheel escapement, is the combination of a toothed gear wheel (giving rise to the *crown* wheel name) and a verge rod that made it possible for the first mechanical clocks to be constructed. This device is critical to the operation of mechanical clocks because it is the means by which the energy stored in weights, or a spring, is controlled by the regular oscillations of a weighted wheel or bar, the ticks, that allow for accurate timekeeping. Later, of course, these regular oscillations would come from the more accurate pendulum, and later still, from the regular oscillations of quartz crystals.

The verge was the only escapement mechanism used in mechanical clocks from their introduction, in the period around 1300, until after the introduction of the pendulum clock in 1656. Although gradually replaced by more accurate escapements, the verge was still found in some clocks even in the late 1800s.

The operation is simple, but also somewhat deceptive. In Fig. 5.3, the crown wheel (C in the figure) is driven by a system of weights, or possibly a spring. It is designed to turn in one direction, and, without any other attachments, would spin out of control, driven by the energy of the weights or spring.[12]

It is prevented from spinning out of control by the verge rod (V), and the two little paddles (p and q in the figure) known as pallets. As the crown wheel moves, it catches on one of the pallets. Because the pallet is attached to the verge rod, which

Fig. 5.3 Verge Escapement

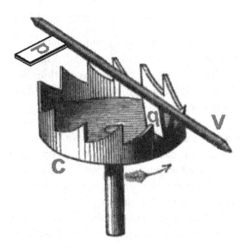

---

[12]Image Credit: Henry Evers (1874), A Handbook of Applied Mechanics, William Collins & Sons, London, Fig. 58, p. 153. Public Domain.

is attached to a weighted wheel or bar (not shown), the crown wheel is momentarily slowed down. As the crown wheel presses against the pallet, it pushes it up, out of the way, and continues to turn. Once again, because the verge rod is attached to a weighted wheel or bar, it swings back down, ready to catch another tooth on the crown wheel. The verge rod, in fact, has two of these pallets, at 90° to each other, for added control.

A simple analogy is to think of the pallets on the verge rod as though they are two swinging trap doors. As the crown wheel turns, it must push its way through one of the trap doors every half a second, or so, making its motion regular and controlled. The exact timing of the tick is controlled by how heavy the trap doors are, and how hard they are to push open. If you want to slow the clock down a little, you make the trap doors heavier, and if you want to speed the clock up a bit, you make the trap doors slightly lighter. Once you have a controlled movement of the crown wheel, you can use this to turn the hands on a clock face. Another analogy is to think of the pallets as speed humps.

The verge escapement is rather like the crankshaft described in *Homo Problematis Solvendis*, in the sense that it converts, or perhaps more accurately modifies, circular motion. The crankshaft, of course, changes circular motion (for example, of a spinning shaft) into reciprocal (up and down) motion (and also converts motion in the reverse sense as well). The verge escapement, on the other hand, *controls* circular motion. We might even argue that the verge escapement mechanism *transforms* the kinetic energy of the spinning wheel into information, and therefore is really the interface between the energy input of the clock, and its information output. Whatever way we label it, this innovation was critical to the development of accurate timekeeping: a problem that has governed much of humankind's development in the last 700 years.

5.3.2 Why Was the Verge Escapement Invented?

Following on from the discussion at the end of the previous section, it seems clear that the immediate problem for which the verge escapement is a solution is the question of *how to control rotation*. Our early Renaissance forebears understood how to transform gravitational potential energy (stored in weights) into the circular motion of gear wheels. This kind of manipulation had been demonstrated, for example, in the Classical Period with the development of the constructions crane in approximately 550 BCE. For timekeeping, however, that motion had to be controlled. Simply letting a gear wheel spin under the action of a weight would provide only a brief burst of movement. Finding a way to spin the gear wheel slowly and with a precise motion made timekeeping possible. Of course, it could be argued that the *real* problem was *how to generate a control signal*. The verge escapement is a solution to both.

5.3.3 How Creative Was the Verge Escapement?

Relevance and Effectiveness: The verge escapement is a delightfully simple solution to the problem of controlling the movement of rotating gear wheel. Readers can see this for themselves by searching online for videos of one of these simple clock mechanisms. Rather than trying to control the movement of a clock mechanism at its source, so to speak, the verge escapement allows an energy source to turn a wheel as well it can, and then applies a very simple braking mechanism to that rotation. While this is somewhat inefficient—it would be a little like allowing your car engine to run at maximum power all the time and controlling your speed only by using brakes—it has the advantage of simplicity.

This innovation correctly reflects the technology available in the era. Our ancestors could fashion crown wheels, verge rods, pallets, and weights with the tools available at the time, and with sufficient precision to make clocks that, while by no means perfect, were acceptably accurate. Sundials were potentially more accurate, but of course could not tell the time in heavy cloud, or at night. Water clocks could work at night, and could also demonstrate good accuracy, but were large and unwieldy, and probably somewhat messy. Mechanical clocks promised a much more versatile, portable and compact device, even if the early versions, with the verge escapement at their core, lost as much as 15 min each day. For these reasons—correct application of knowledge and techniques of the day, reasonable performance and an acceptable adherence to the constraint of accuracy—I will give the verge escapement 3.5 out of 4 for this criterion.

Novelty: The verge escapement highlights the key to accurate timekeeping, and this has remained the core of modern clocks. The creation of an accurate control signal—whether the punctuated movement of a crown wheel, the precise oscillation of a pendulum, or the rapid vibration of a quartz crystal—is what allows mechanical, and now electronic clocks to keep time. The verge escapement emerged because of the limitations of previous devices (sundials and water clocks, for example), and, although less accurate to begin with, showed our ancestors that they were on the right path.

By departing from previous methods that relied largely on naturally occurring phenomena (the movement of the Earth around the Sun, or the action of water under gravity), mechanical clocks, with the verge escapement at their heart, opened up a new field of artificial, mechanical timekeeping, with enough promise to entice our ancestors to pursue incremental improvements and occasionally, wholly new sources of the underpinning control signal. I give the verge escapement 3.5 out of 4 for the criterion of novelty, to reflect these qualities. The concept of timekeeping was not new, but the way that the verge escapement performed the core function was new and would have surprised our Renaissance mothers and fathers.

Elegance: To our modern eyes, mechanical clocks (and their verge escapements) of the fourteenth century may look rather clumsy, over-sized, and even crudely made. However, we must always endeavour to view these innovations through the eyes of someone experiencing them at the time of their development. Imagine a young woman entering Salisbury Cathedral, in the South of England, on a Sunday morning in 1390.[13] When told that the strange arrangement of large iron gear wheels, ropes, weights and, of course, the oscillating verge escapement mechanism, were responsible for calling her to Mass that morning, she must have marvelled at the technology! To someone with a degree of technical knowledge, perhaps a blacksmith, used to working iron, the mechanism would have been an impressive sight, with skilfully executed gears, and a well-proportioned, harmonious structure. Perhaps the only slight blemish might have been its limited functionality—it is thought that the clock only struck a single bell, four times a day, to signal Mass—and what may have been a rather tedious, frequent and laborious winding mechanism (to raise the weights that powered it). However, keeping in mind that our analysis is for the verge escapement, and not the overall clock (although one is almost inseparable from the other), these factors detract only slightly from what remains an impressively elegant design. I can imagine its creator, our unknown *Femina Problematis Solvendis*, explaining the function of the verge escapement to someone else, and this person immediately understanding how the pallets slow down the crown wheel: the mechanism is readily understandable! I give the verge escapement itself a score of 3.5 out of 4 for elegance.

Genesis: If novelty is fundamentally inward looking—how new is the invention in a narrow sense—then genesis is about the newness of the artefact with respect to the wider field of endeavour. Genesis is about the extent to which an invention is a paradigm-breaker. Novel innovations are often more incremental in nature, reflecting advances on known pathways, but genesis asks to what extent an invention breaks entirely new ground. It seems hard to go past a strong score for the verge escapement in this category. It was essential to the *creation* of mechanical clocks, in the same way that pendulums were essential to *accurate* mechanical clocks. I like to think of incremental improvements—novelty in the narrower sense—as a case of moving along a curve of increasing returns. Incremental changes deliver improved performance, at least until the point of diminishing returns is reached. Beyond this point, however, the only option is for paradigm change. Water clocks, sun dials, hour glasses, and the like all had their benefits, but timekeeping was reaching, and possibly had reached, its point of diminishing returns. This provided the impetus for an astute woman to look for a new paradigm. The verge escapement was such an invention. It established a new baseline for timekeeping and drew attention to the weaknesses of other methods. It reconceptualised the measurement of time, and, while this mechanism may have been limited in its wider application and would be superseded by even better control signal mechanisms, the verge escapement must be given a score of 3.5 out of 4 for genesis.

Total: The verge escapement achieves a very creditable score of 14 out of 16 on our creativity scale. This places it just inside our *very high* band and compares favourably to the other inventions of the early to mid-Renaissance that we have considered, both in this volume, and in *Homo Problematis Solvendis*.

---

[13]Salisbury Cathedral, about 150 km South-West of London, claims to have the world's oldest functioning mechanical clock. It was used to control the striking of bells to signal Mass.

Chapter 6
The Renaissance (1300–1700): Reawakening Creativity

We follow the Dark Ages, allowing a little bit of overlap, with the period known as the Renaissance. This time of *rebirth* began with a renewed interest in learning, drawing on classical sources, and reforms in the process of education. Another defining characteristic of the Renaissance was the development of linear perspective (in the artistic sense), which was part of a more general move towards greater realism in painting. A general surge of interest in observation of the world, and the development of inductive reasoning—formulating explanatory theories based on observations of the world—were organised for the first time in a more formal way, and *experimentation* was an innovation in how humans made sense of the world in which they lived.

The first innovation that we consider in this period is central to one of the drivers of this book. *Feminism*, the equal treatment of both women and men, is not something that our ancestors should have had to invent. Nevertheless, as humans transitioned from the nomadic, and egalitarian, hunter-gatherer lifestyle, to a more sedentary, agricultural way of life, it seems that *inequality* became the norm. This was entrenched in societies like those of Rome and ancient Greece: women could not vote, and they had many other restrictions on their lives. Even though women still wielded influence, and even power, sometimes in their own right, or sometimes through their husbands and sons, this was the exception rather than the rule. What little power and influence women did have was also unevenly distributed, with wealthy women faring the best in this respect. The slow process of unravelling this inequality began, in 1405, with Christine de Pizan's novel thinking.

Our second invention in this period is far more mundane than Feminism, but one that we all use, twice a day if we are doing things right. The *Toothbrush* was invented in around 1498 in China and has changed remarkably little in the intervening 500 years. If you've ever had a toothache, you don't need me to explain the benefits of a simple, preventative routine like brushing. The lack of pain killers and modern dentistry must also have been a powerful motivator for this invention.

© Springer Nature Singapore Pte Ltd. 2020
D. H. Cropley, *Femina Problematis Solvendis—Problem solving Woman*,
https://doi.org/10.1007/978-981-15-3967-1_6

Finally, we close out the Renaissance with another medical invention. You could say that this invention—*Palliative Medicine*—is what happens if you don't brush your teeth. The innovation here was the first systematic attempt to relieve chronic pain, as well as to ease the suffering of those with terminal, painful conditions. This also reflected a change in our ancestors' understanding of disease and pain. These were not things entirely beyond their control, to be suffered and tolerated. Our growing understanding of the world allowed our ancestors to begin to fight back against some of the mysteries of life.

6.1 Feminism (1405)

The man or the woman in whom resides greater virtue is the higher; neither the loftiness nor the lowliness of a person lies in the body according to the sex but in the perfection of conduct and virtues. (from The Book of the City of Ladies*)—Christine de Pizan, Venetian Writer and Feminist (1364–c1430)*

Our first Renaissance innovation[1] is a central theme in Christine de Pizan's composition, *The Book of the City of Ladies*, from which the quotation that begins this section is drawn. *Feminism*, in the uncomplicated, *dictionary-definition* sense of the word, is simply the goal of achieving equality of the sexes. Over many decades, in modern times, this has sometimes been associated with a range of connotations, not all of which are positive. However, leaving aside any discussion of the means and the different interpretations that have been applied to the term, the desired end is hard to argue against. While opponents might frequently suggest otherwise, equality—whether of race, income, or gender—is *not* a zero-sum game. What this means is that there is *not* a finite amount of equality to be distributed, just as there is not a finite amount of human creativity.

Emma Lazarus, the nineteenth century American writer and activist said that "Until we are all free, we are none of us free", and the same applies to gender (and other forms of) equality. A world in which some people have more rights than others, is a world in which no one's rights are secure. Such a world implies that these rights can be taken away, transferred or traded. A world in which we are all free, and equal, by contrast, is a world in which the *differences* have no value. Equality of the sexes removes the market value, and the incentives, from differences.

---

[1] I have a confession to make here. In the first draft of this book, I had *Feminism* featured as the last invention in the preceding epoch, the *Dark Ages*. For some reason, and because the Dark Ages overlaps with the Renaissance, I thought that Feminism was a great way to round off the former, rather than kick off the era defined by a reawakening, and a transition from unenlightened, medieval thinking to a more modern view of the world. I see now that this was a dumb idea, and I am glad that I moved it!

6.1.1 What Was Invented?

Feminism, of course, is not a modern invention, although the term itself is more recent. For as long as women have been treated differently in political, economic, social, and/or personal terms, there have been women (and some men) who have fought to rectify this imbalance. We begin our chapter on the Renaissance by examining one of these women: Christine de Pizan (1364–c1430), and her invention of the concept that women and men should be treated equally. Although it may seem strange to call it this, we can classify feminism as an information-handling system.

Christine de Pizan was born in Venice, the daughter of a Councillor of the Republic of Venice. When she was only four, her father accepted an appointment, as astrologer, to the court of France's King Charles V in Paris. She lived well, marrying a member of the King's court. Tragedy struck, however, when she was only 24. Her father died of the plague in 1388, and the following year she lost her husband to the same disease. This left her suddenly responsible for three children and the care of her mother. Worse was to come, with legal problems denying her access to money from her husband's estate. Critical to our discussion is the fact that she rose to the challenges she faced, becoming a prolific and successful writer. She began by writing love ballads which attracted the attention of wealthy patrons. As she secured herself in her occupation as a writer, she turned to a more diverse range of styles and topics, including political treatises.

Our special focus here is Christine de Pizan's book, written in 1405, called *The Book of the City of Ladies* (Fig. 6.1 depicts an image from the book). In this work,

Fig. 6.1 Master of the City of Ladies Christine de Pizan

she created a fictitious city, populated only by women, including three allegorical characters: Lady Reason, Lady Justice, and Lady Rectitude. Through these characters (depicted on the left of Fig. 6.1, in discussion with Christine de Pizan herself), she was able to construct a dialogue in the form of questions and answers. Critically, this dialogue addressed matters from a female perspective, allowing Christine de Pizan to explore *feminist* issues and concerns. *The Book of the City of Ladies* was followed, in the same year, by *The Treasure of the City of Ladies*. Written using a similar allegorical device, the second volume extended her exploration of this female society.[2]

Christine de Pizan was acknowledged, in her own lifetime, as the first professional female writer in Europe. Her works, including those concerning the City of Ladies, remained in print for over one hundred and fifty years. Elizabeth I of England, who reigned from 1558 to 1603, was known to have copies of several of de Pizan's works, including *The Book of the City of Ladies*. What stands out in the work of Christine de Pizan is her vigorous defence of women, including outspoken rebuttals of popular misogynistic texts of the time. Although the label *feminism* may be a modern construct, the spirit and philosophy embodied in feminism is abundantly apparent in de Pizan's work and seems an appropriate introduction to the Renaissance Era.

6.1.2   Why Was Feminism Invented?

As I have already indicated, Christine de Pizan may not have called herself a feminist, but her arguments and ideas clearly supported both the broader notion of equality more generally, and the more specific concept of gender equality. I am not going to attempt to delve into deeper philosophical territory on this subject, not least because the world doesn't need me mansplaining feminism! Suffice it to say that Christine de Pizan is widely acknowledged for her work, and her courage, in addressing these matters. What problem was she tackling, for which her ideas were the solution? Perhaps *how to establish equality*?

6.1.3   How Creative Was Feminism?

Relevance and Effectiveness: If I have upset anyone in this case study (and I sincerely hope I have not), then I am going to double-down and give Feminism a score of only 2 out of 4 for relevance and effectiveness. However, hear me out! It has been an important prerequisite of our discussions on creativity that inventions must work. I

---

[2]This image is an illustration from the book of the city of ladies. Image Credit: The yorck project (2002) 10,000 *Meisterwerke der Malerei* (DVD-ROM), distributed by DIRECTMEDIA Publishing GmbH. Original document is held in the Bibliothèque Nationale de France. Public Domain.

will explain, later in this book, that Leonardo da Vinci, in my view, cannot really be called an inventor for this reason. A similar problem faces Christine de Pizan's innovation. Equality of the sexes was certainly not achieved in her lifetime, and indeed, is still some way from being a *fait accompli* even today. Unfortunately, this means that we cannot be too generous in our analysis, as far as *performance* is concerned. What is important for our discussion of creativity is *why*? Was the notion of gender equality simply too much for many people at the time? Did it exceed conventional knowledge of the time, meaning that it was far ahead of its time? In a just world this should not have been the case, but the reality is that the society in which Christine de Pizan lived was not yet ready for gender equality.

Novelty: In stark contrast to the previous category, Feminism must score the maximum for novelty. Indeed, that fact that it was so strong in drawing attention to the shortcomings of a lack of equality (*diagnosis*), and so powerful in showing how the status of women that existed at the time could be improved (*prescription*), means that it really cannot be given anything less than a 4 for novelty. *The Book of the City of Ladies* also set the scene for further refinements to gender equality and feminism and was unequivocally a radically new approach.

Elegance: I normally link elegance with effectiveness, usually noting that elegance is not just coincident with effectiveness but may be a *cause* of effectiveness. Equally, poor elegance can drag down effectiveness. However, in the case of feminism, I think it is unfair to put the blame for the low effectives score on poor execution. We could probably make some criticisms of de Pizan's convincingness, for example, in the literary style she uses to make her points. Clearly, we might be able to argue that some people (presumably, men) did not find her ideas pleasing, or her arguments complete and fully worked out. We might even argue that her book has inconsistencies and is therefore somewhat lacking in gracefulness. All of these might be the justification for a score of 3 out of 4, but none of these are the reason for the low effectiveness. That seems to come down to the recalcitrance of society, rather than the weakness of the ideas!

Genesis: Like novelty, the genesis of Christine de Pizan's feminism must be high. Here, we see strong elements of a paradigm shift. This first foray into gender equality sparks ideas for how the concept might be developed further (*foundationality*) and suggests new ways of looking at other problems (e.g. racial equality). The invention draws attention to previously unnoticed problems and sets a new benchmark for how people consider what it means to be equal. There can be little doubt that feminism deserves 4 out of 4 for genesis.

Total: Christine de Pizan's invention—that women should be treated as the equal of men—scores a *very high* 13 out of 16. I have already spoken about what I think is the underlying cause of the weakness in effectiveness, and potential issues with elegance. There is probably little more that Christine de Pizan could have done to improve her invention. Someone had to take a deep breath and let the genie out of the bottle, even if it would take generations for feminism to effect real change.

6.2   The Toothbrush (c1498)

Treat your password like your toothbrush. Don't let anybody else use it, and get a new one every six months—Clifford Stoll, American Astronomer and Author (1950–)

Our modern human ancestors, at least as long ago as 3000 BCE, obviously understood the desirability of keeping their teeth clean. Archaeological evidence, from both ancient Babylonian and Egyptian societies, tells us that our forebears would use twigs, which they first chewed on to fray the ends, to make a simple kind of brush. If you have ever had a piece of food stuck between your teeth, you have probably also used a toothpick, or a piece of floss, to remove it, and we can imagine our ancestors using whatever material might be to hand—a shard of bone, a woollen thread—to perform a similar function.

Humans, of course, are nothing if not creatures of habit, and it is easy to picture our ancestors carefully keeping a small shard of bone or sliver of wood that made their favourite toothpick. It is also not hard to imagine some enterprising mother, perhaps tired of dealing with the consequences of her children's toothaches, looking for a more reliable solution to the problem of dental hygiene.

This is far from a minor problem. Even today, in a modern society like Australia, with a population of about 23 million, the cost of poor dental health is surprisingly large. It was estimated that, in 2008/09, hospital admissions for dental disease in Australia cost the nation A\$84 million (about US\$50 million)!

Remember that physiological needs sit at the base of Maslow's hierarchy, and that a healthy child is, to some degree at least, a safe child. If you have ever had a really bad headache, whether from a decayed tooth, or something else, you also know how debilitating this can be. This must have been a powerful driving force for history's mothers. It seems almost inevitable, therefore, that at some point in history, an enterprising, inventive woman would fashion an improved brush from more effective and robust materials and give us the first toothbrush in the form that we still use today. The toothbrush is a simple material-handling system—literally designed to remove *material* from the surface of our teeth—and we will now examine this simple invention in more detail.

6.2.1   What Was Invented?

Like many other important inventions (e.g. paper—see *Homo Problematis Solvendis*), it was the Chinese who take the credit for the invention of the toothbrush. The date we have used—1498—is merely the date on which Europeans, emerging from their Dark Ages, discovered what had already been invented elsewhere.

The toothbrush of this era hardly needs a detailed description. As readers can see from the example depicted in Fig. 6.2, it is remarkably similar to those we are familiar with today. The brush has a long, palm-sized handle that is shaped or textured to improve the user's grip. The head, then as now, must fit into the user's mouth, and

Fig. 6.2 The Wood and Bristle Toothbrush

must be shaped in such a way that it doesn't scratch our mouths, tongues and gums. It must be big enough to serve as an anchor for the vital bristles, but not so big that it cannot reach our back teeth.[3]

In 1498, although superficially almost identical to our modern toothbrushes, the device itself was typically made from wood, bamboo or bone and the coarse bristles from a boar. We can also imagine that toothbrushes in this period were something of a status-symbol. Not that people would parade around the streets brandishing their toothbrushes to the shame of others, but in the sense that only wealthier people had the means to obtain them. Unlike modern, mass-produced, plastic brushes, a bone toothbrush, for example, would have required the services of a skilled artisan. Toothbrushes were therefore custom-made in this era, and therefore must have been beyond the means of many ordinary people.

There is one other striking feature of the toothbrush from this Renaissance Period of history. Because of its close similarity to its modern descendant, it is possible to make some observations about the nature of incremental innovation. In a period of over 400 years, only three things have changed about the toothbrush, and all of these have simply become better or cheaper. The material used to construct the handle/head is now typically plastic instead of wood or bone. The bristles are now also usually a plastic material rather than boar bristles. The shape has changed a little, with a smaller head, and an angled design to make it easier to reach back teeth. The point is that none of these changes is radical in nature. The paradigm remains unchanged, and a point of diminishing returns has not yet been reached. Of course, this does not mean that dental hygiene has not changed radically, but it is interesting that one key element of our dental health has remained almost untouched for at least 400 years! Does that mean that the toothbrush peaked early? We will get some insight into this when we assess its creativity.

---

[3]Image Credit: https://toothcorner.com/brushing-toothbrush-history/.

6.2.2   Why Was the Toothbrush Invented?

For once we get an easy problem to define! It seems to be very clear that the problem the toothbrush was developed to solve was, very simply, *how to clean teeth.*

6.2.3   How Creative Was the Toothbrush?

Relevance and Effectiveness: My previous comments about the toothbrush peaking early and changing very little over the last 400 years may have lured you into thinking that this criterion will be a high score. If the design is largely unchanged in 400 years, then surely the toothbrush is very effective? However, I am giving it only a 3 for relevance and effectiveness! The issue here is that although it has some excellent qualities—it fits very well within the constraints of size, for example—and it is also technically and materially a good solution, it has a question mark over basic performance. It certainly assists in cleaning teeth, but just how well is, as it remains today, somewhat variable. Even with modern versions of the toothbrush, it is possible to miss nooks and crannies, and your dentist, like mine, probably still urges you to floss your teeth as well as brushing. Both this 400-year-old version, and its modern equivalent clean teeth *fairly well*, but not perfectly. If they did, none of us would ever need to visit the dentist again. Therefore, in terms of function and performance, it was, and remains, imperfect and receives 3 out of 4!

Novelty: This criterion requires us to consider both how the invention helps us define the problem we are solving, and, to what extent the invention sheds new light on that problem. In the former case (problematisation) the toothbrush is a strong contender. Imagine being given one for the first time as a replacement for your chewed twig. We can imagine that it would have been immediately obvious just how much better this device was and highlights the nature and benefit of simple improvements to the brushing concept. In the case of the latter issue (propulsion), the toothbrush is probably a little weaker. It is a modest extension of the chewed twig (which is really just a very crude brush). It is not a *radically* new approach to what preceded it, and it doesn't really change the paradigm. It is an improvement over what went before, but we can imagine that it would not have blown people's minds. It was simply a fancier version of the chewed twig! So, while it has some important qualities in this category, the very incremental nature of this invention means that we can give it only a 3 out of 4 for novelty.

Elegance: In contrast to the previous two categories, elegance is very strong for the toothbrush. Once again, imagine being given one as a replacement for your trusty old chewed twig. The toothbrush of 1498 must have seemed like an impossibly modern and high-tech leap forward (even though, as we have discussed, it was really only incremental in nature). It was well worked-out, well-proportioned, gracefully executed, and pleasing to the eye. It must have been a pleasure to own and use one rather than a grubby, and probably unpleasant-tasting stick! It is therefore easy to give it a maximum score of 4 for elegance. It is interesting to note, however, that high elegance has not, in this case, resulted in high effectiveness. Elegance in this case was perhaps a little more exclusively aesthetic in nature.

Genesis: The toothbrush is turning into a rather curious innovation. Did it change how the problem—how to clean teeth—was understood? Not really. It was still conceived as a matter of sweeping dirt (plaque, food particles) from the surface of the teeth. The chewed twig attempted to do this, and the modern toothbrush still targets the same issue. The toothbrush did not establish a new paradigm, even if it did set a new benchmark for the existing approach. What I find a little surprising, in retrospect, is that a key element of modern dental hygiene remains this somewhat flawed device. That doesn't mean that we should abandon the toothbrush. What surprises me more is that we have not really developed a new tooth-cleaning paradigm, even after 400 years. It is true that we have electric toothbrushes, but even these are just squeezing a little more value out of a 400-year-old paradigm. Perhaps this is just a reflection of the fact that the toothbrush is good enough for the problem, and the improvements are not to the device itself, but to other aspects of dental hygiene (floss, fluoride, fillings, and so on). The more I think about it, the more curious I find this. I give the toothbrush a score of 2.5 out of 4 for genesis.

Total: The toothbrush ends up with a total of 12.5 out of 16. As with many of our inventions, both in this volume, and in *Homo Problematis Solvendis*, this score places the innovation in the region between *high* and *very high* creativity. Somewhat unusually, it scores less than the maximum for effectiveness, especially given its high elegance. It surprises me that 400 years of development have probably only managed to find another 0.5 for effectiveness (i.e. still not perfect), and have only made a cheaper, nicer-looking version of the same old device. The question is, what is the paradigm-shattering innovation in dental hygiene that will render the toothbrush obsolete?

6.3 Palliative Medicine (c1570)

Death does not take the wise man by surprise, he is always prepared to leave—Jean de la Fontaine, French Poet (1621–1695)

If you are keeping track, you will have noticed that only three of our inventions so far (out of 12) have a known (and female) inventor. At the same point in the previous volume (*Homo Problematis Solvendis*, which had an unintended male bias) the proportion was, in fact, the same. This is perhaps not surprising, given that prehistory, by definition, has no record of the actual inventor, while subsequent epochs (the Classical Era, the Dark Ages) often leave us with incomplete descriptions and records. Happily, whether male or female, it is around the current period— roughly speaking, the middle of the Renaissance—that history begins to record, more rigorously and systematically, just who invented what. Although women would often not receive credit for their inventions until much later than this, for a variety of unfair reasons, I am reassured by the fact that this phenomenon was not as long-standing or pervasive as I feared.

So, we turn now to the work of Loredana Marcello, a Dogaressa[4] of Venice, from the time of her husband's election to the role of Doge in 1570, until her death in 1572. Prior to her time as Dogaressa, she was already known as a highly erudite person, renowned as an exemplar of the educated Renaissance women. We study her now for her work in the development of medicines associated with the plague, stemming from her study of botany.

6.3.1 What Was Invented?

Loredana Marcello is believed to have invented what we would probably now think of as *herbal treatments* for the plague. We know that she studied botany, and also that she had studied under Melchoirre Giuliandino, a professor at the Botanical Garden[5] in Padua. Unfortunately, specific details of these treatments are lost to us, but we can draw some inferences based on the little we know of Marcello and the circumstances of the time.

The first piece of relevant information, beyond her education in medicinal botany, was that Europe had experienced a continent-wide outbreak of the bubonic plague between 1346 and 1353. It is thought that this may have killed up to 60% of the population of Europe, causing massive social and economic upheavals. Indeed, estimates suggest that it took some 200 years for the population of Europe to recover to prepandemic levels. Venice had not been spared this outbreak, and in fact, experienced a further 22 outbreaks of bubonic plague between the years 1361 and 1528. The point here is that the *Black Death*, as the fourteenth century pandemic has become known, must have loomed large in the memories of cities across Europe. It was an ever-present threat, lurking in the shadows, with little understanding of the cause, let alone any knowledge of a cure. The reason this helps us understand Loredana Marcello is that anyone who studied botany in this period, and at the Botanic Garden in Padua, was probably interested in the plague.

The second piece of relevant information is that we now know what the cure is—it is antibiotics—and these, of course, were not invented until the twentieth century (see *Homo Problematis Solvendis*, and the discussion of the invention of *practical* antibiotics by Howard Florey and Ernest Chain).

---

[4]In Renaissance Venice, the *Dogaressa* was the spouse of the *Doge*. The Doge, for his part, was the senior elected official of the Republic of Venice (726–1797)—chosen by the city-state's aristocracy—and was chief magistrate and leader. Dogaressas, although wielding no political power in law, in practice were frequently influential in their own right, much like the First Lady in US politics.

[5]The *Orto Botanico*, in Padua, Italy, was the world's first purpose-built botanical garden, and was established in 1545 by the Venetian Republic. It was established specifically for the growth of medicinal plants, and as a place where students (of the University of Padua) could learn to distinguish these from non-medicinal varieties.

This brings us to Loredana Marcello's innovation. What is the purpose of medicines for diseases and conditions that have no cure and have high mortality rates (typically 50–70% for the plague)? Both for the majority who would succumb to this disease, as well as the lucky minority who survived, the bubonic plague involved painful swelling of lymph nodes, nausea, vomiting, fever, internal bleeding, and gangrene. The best that could be offered in the days before effective antibiotic treatments was to hope to ease the pain and relieve the symptoms of the sufferers. Loredana Marcello developed herbal treatments that must have been directed towards achieving these goals and she is therefore associated with the creation of the formal discipline of *palliative medicine* (see Fig. 6.3).[6]

Fig. 6.3 Saint Carlo Borromeo ministering to the plague victims

The origins of a palliative approach to terminal illnesses have been linked with Sir Francis Bacon who, in 1605 "challenged physicians to accept the new responsibility of easing the pangs of dying persons so they could expire with greater tranquillity[7]". Thus, although explicitly stated in 1605, Loredana Marcello seems to have anticipated this approach to terminal illnesses by some 35 years. Because the bubonic plague could not be cured in this era, the focus of Loredana's treatments must have been intended as palliatives. What this means is that her work relieving the suffering of patients was not an accident: she intended to relieve their suffering, knowing that she could not cure them. For this reason, we can confidently credit her with the invention of this approach to terminal disease.

Palliative medicine, in its modern sense, is a much more complex, interdisciplinary specialisation focused on the relief of symptoms, pain and also the related physical and psychological impacts of chronic, i.e. long-lasting, medical conditions. It has also been described, in the past, as *hospice care*, and in that sense is associated more with the care of patients with terminal illnesses.

6.3.2 Why Was Palliative Medicine Invented?

Our *how to verb noun* challenge, in this case, seems to be simple. Loredana Marcello set out to relieve the pain and suffering associated with the bubonic plague. Because many of the symptoms of the plague are not unique to that disease—pain and nausea, for example—it must have been apparent to her that her herbal treatments were also valuable for easing symptoms of other chronic and terminal diseases. Therefore, we can define the problem she set out to solve as *how to relieve symptoms*. If we wish to constrain this a little and keep the focus on the plague, then we can modify our statement slightly to *how to relieve plague symptoms*.

6.3.3 How Creative Was Palliative Medicine?

Relevance and Effectiveness: Loredana Marcello's innovation is probably the toughest one to score in this book. We do not know the full details of her treatments, forcing us to make some educated guesses. There appear to be no records that can tell us how many people received her treatments, or their impact. Did patients who received Marcello's herbal medicines suffer less than others? Did her treatments prolong their lives? Did a higher than normal proportion of her patients survive the devastating effects of bubonic plague? We will almost certainly never know. However, the fact

---

[6]Image Credit: Oil painting by Pierre Mignard the Elder (1612—1695). Source: https://wellcomecollection.org/works/zym6q8t3. creative commons 4.0 https://creativecommons.org/licenses/by/4.0/legalcode.

[7]From the 2006 book *Palliative Care: The 400-Year Quest for a Good Death*, by Harold Vanderpool.

that we know of Loredana Marcello today suggests that her innovation must have had some noteworthy effect. Herbal medicines did not exceed the knowledge available in this era, and her treatments must have been at least somewhat effective in relieving plague symptoms (or in relieving some symptoms of the plague quite well). For these reasons, we can justify a score of 2.5 out of 4. Anything less than 2 and we would have to say that it was, on balance, not effective. Something over 2 suggests at least moderate effectiveness.

Novelty: When we judge Marcello's innovation in palliative treatments in terms of novelty, she fares somewhat better. If Francis Bacon is credited with the notion of easing the pain and suffering of the dying, in 1605, then Loredana Marcello was already thinking along these lines some decades earlier. Her attempts to relieve plague symptoms helped to clarify what the problem was—how to *relieve* symptoms. She reframed the problem, and disease was no longer a matter completely beyond human control. Even if it could not be prevented or cured, it could, at least, be controlled in some way. Steps could be taken to improve previous approaches. More could be done for the terminally ill than just mop their brows and keep them warm. Marcello shed new light on the whole problem of disease and the symptoms of disease. She took an early, and important, step along a path whereby our ancestors would eventually cease to see disease as incomprehensible and mysterious. For these reasons, although possibly of doubtful effectiveness, Marcello's palliative medicines deserve a score of 3 out of 4 for novelty.

Elegance: Time and again we are seeing a basic truth in the assessment of the creativity of inventions. Effective innovations work well, in part, because they are well-executed, complete and understandable. If an invention is well-made, the chances are, it will work well. Conversely, poor execution, or inventions that are incomplete, or ill-proportioned, frequently don't have strong effectiveness. It is not that the poor elegance—which is what we are talking about—is deliberate. Rather, weaker elegance usually seems to be associated with external limitations. The inventor might have an idea that is difficult to execute with the materials or technologies available. This seems to be applicable to Loredana Marcello's palliatives. If we are right, and they were not fully effective (remember that we gave them a score of 2.5 for effectiveness), it is not hard to make a case that this might have been due, to some degree, to the tools and technologies of the day. Making medicines is not the same as cooking, even though both can be thought of as similar activities. You might get away with rough measurements of flour and water when making pastry for a pie, but when making a medicine, with smaller quantities of more delicate active ingredients, greater accuracy would be important. Similarly, testing and documenting the effects of different recipes would be difficult without the kinds of modern measuring instruments that we now possess. I give Marcello's palliatives as score of 2.5 out of 4 for elegance. Like effectiveness, they must have had a modest degree of good execution, but not enough, thanks to technological limitations, to support better effectiveness.

Genesis: In this category, we are back to a stronger view of Loredana Marcello's invention of palliative treatments for the plague. If novelty helped to define the problem, and shed new light on the problem, then genesis fundamentally altered our forebears' understanding of the problem. Under genesis, we can see that the introduction of palliative treatments changed the *philosophy* of medicine. Palliatives were not merely the logical extension of earlier medicines but tried to do something quite different. In other words, the paradigm shifted from trying (unsuccessfully in the case of the plague) to cure the disease, to trying to manage the disease. We must, however, accept that Loredana Marcello's approach was the first step on this new pathway. The paradigm shift would solidify and develop further, but this innovation deserves a score of 3 out of 4 for genesis.

Total: Loredana Marcello's invention of palliative treatments for the plague scores 11 out of 16 for creativity. This places it in the *high* range for creativity, although it is one of the lowest in our catalogue of *Femina* inventions. Its principal weaknesses lie in effectiveness and elegance. With better methods, instruments, and tools, bringing better execution and therefore greater elegance, it is likely that these treatments would have improved, and we would be analysing a score of 13–14. Despite being one of our lower scoring innovations, Marcello's palliatives played an important role in changing medical thinking and must therefore still be admired for their particular creative strengths.

Chapter 7
The Age of Exploration (1490–1700): New Worlds and New Problems

The Age of Exploration, also referred to as the Age of Discovery, was a period during which European societies began to look outwards. The rebirth of knowledge, and humankind's interest in its own place in the world, marked by the Renaissance, set the stage for the application much of what had been learned since about 1300. The Age of Exploration, covering the time from the end of fifteenth century to the close of the seventeenth century, saw many great geographical discoveries. Although this spirit of exploration was not entirely new (think of Marco Polo's journey to the Far East between 1271 and 1295), Christopher Columbus sailed West in 1492, ultimately opening up both South and North America to settlement by Europeans. Ferdinand Magellan led the first expedition to circumnavigate the Earth in the early 1500s, and the great Southern continent (Australia) was first sighted by Abel Tasman in the 1640s.

Not surprisingly, this surge in exploration and discovery proceeded in parallel with a surge in invention. Exploration demanded accurate navigation, and accurate navigation needed better methods for telling the time and for calculating the positions of celestial objects. Therefore, our first invention in this era is an accurate and functional set of *Astronomical Tables*, vital for helping sailors to find their position by the stars and planets.

The second innovation we consider, however, is a complete contrast to navigation and travel. With the *two-handed internal rotation,* we return to another ever-present theme of human innovation: medicine. Like palliative medicine in the preceding era, this invention reflected a growing understanding of the internal, unseen world. In earlier time periods, humans may have felt that they were completely subject to the whims of the Gods when it came to questions of childbirth. However, our ancestors in the seventeenth century understood that they could do things to influence the outcomes of these phenomena. Childbirth has always been a difficult and potentially dangerous process, but Justine Siegemund helped to dispel some of the mystery and danger with techniques that could correct so-called malpresentations.

© Springer Nature Singapore Pte Ltd. 2020
D. H. Cropley, *Femina Problematis Solvendis—Problem solving Woman*,
https://doi.org/10.1007/978-981-15-3967-1_7

Our final innovation in this epoch is neither astronomical nor medical, but an excellent example of our ancestors' growing prowess in technology. If Justine Siegemund was expert at manipulating the *internal world*, then Sybilla Masters, with her *Corn Mill*, showed us how to harness the external world—in this case water power—for the benefit of society.

7.1　Astronomical Tables (1650)

It's not that we have little time, but more that we waste a good deal of it—Seneca, Roman Philosopher (4 BCE–65 CE)

In *Homo Problematis Solvendis*, I discussed several inventions that are all tied together by the desire of humans to travel. Thus, the oar, the slide rule, the pendulum clock, the velocipede, the Wright Flyer, and even Sputnik all tie, in some way, to needs that end up focused on our ability to move. Whether that ability is to move faster, farther, with greater accuracy, or more efficiently, it seems significant that many of our most important inventions are underpinned by a need to *move*. Some of these inventions have, not surprisingly, been energy-handling systems, while others are focused on information-handling. The invention we consider now was an example of one of the great information-handling problems that have occupied humankind's quest for movement for millennia: how do we know where we are, and where we are going? Even our ancient ancestors understood that the night sky held the key to solving these problems, but it would take precise, and accessible, astronomical tables (coupled, of course, with accurate clocks) to help transform theory into practice.

7.1.1　What Was Invented?

Maria Cunitz (1610–1664) was a Silesian[1] astronomer, regarded as the preeminent female in this field in her day. Maria was born into a well-to-do and educated family of German heritage. She was known to be able to speak seven languages—not only major European languages but also Latin, Greek, and Hebrew—and she was a musician and painter, as well as having a special ability in complex mathematics. It was this mathematical ability that led to our present focus, her publication of the book *Urania Propitia* (Fig. 7.1). Cunitz's innovation was the development of a set of astronomical tables: a simplified and more user-friendly version of those published as the *Rudolphine Tables* by Johannes Kepler in 1627.[2]

---

[1] Silesia is a geographic region in central Europe. Nowadays located mainly in Poland, the *nationality* of the region has changed many times, for example, moving out of a long period of German control as a result of the Potsdam Agreement that settled territorial issues at the end of World War 2.

[2] Image Credit: Maria Cunitz. Printer: Johann Seyfert. First published 1650. Public Domain.

URANIA
PROPITIA

SIVE

Tabulæ Aſtronomicæ mirè faciles, vim
hypotheſium phyſicarum à Kepplero pro-
ditarum complexæ; facillimo calculandi compendio,
ſine ullâ Logarithmorum mentione, phæno-
menis ſatisfacientes.

Quarum uſum pro tempore præſente,
exacto, & futuro, (accedente inſuper facillimâ Superio-
rum SATURNI & JOVIS ad exactiorem, & cœlo ſatis conſonam
rationem, reductione) duplici idiomate, Latino & vernaculo
ſuccinctè præſcriptum cum Artis Cultoribus
communicat

MARIA CUNITIA.

Das iſt:

Newe und Langgewünſchete/leichte

Aſtronomiſche Tabelln/

durch derer vermittelung, auff eine ſonders
behende Arth/ aller Planeten Bewegung/nach der länge/
breite/ und andern Zufällen/auff alle vergangene/gegenwertige/ und künfftige Zeit-
Puncten fürgeſtellet wird. Den Kunſtliebenden Deutſcher Nation zu gut/
verfertiget.

Sub ſingularibus Privilegiis perpetuis,
ſumptibus Autoris, BICINI Silurorum.,

Excudebat Typographus Olſnenſis JOHANN. SEYFFERTUS,
ANNO M. DC L.

Fig. 7.1 Cover page from *Urania Propitia* (1650)

Maria Cunitz's work can be regarded as an information-handling system and can
be divided into three sections. The first is a set of *tables for spherical astronomy*.
This branch of astronomy is reckoned to be the oldest and is concerned with the
positioning of objects in the celestial sphere as seen from a given location on Earth,
at any specified time and date. The basis of celestial or *astro*-navigation—think of
sailors finding their positions on a chart using a sextant—spherical astronomy was
critical not only for navigation but also for religious and astrological purposes. Tables
for spherical astronomy would specify the positions of celestial objects in the sky.

A sailor, using these tables for navigation would, in very simplified terms, compare the difference between the position of, say, Saturn, as seen from a known location, with the actual position of that planet measured at the sailor's location. Using the difference between these two values, and repeating these measurements for several different celestial objects, the sailor could work out how far away from the known location she was, and thereby fix her exact position.

The second section of Cunitz's book consisted of a set of *tables of average motions*. Maria Cunitz's specific achievement here was again a simplification of Kepler's calculations, in this case, for determining the positions of the solar system's planets within their orbits. Where should Mercury, for example, be on any given day, at any given time? What about Venus or Jupiter? Of course, we know approximately where these planets should be, because they are confined to the so-called *ecliptic*. They orbit the Sun on a fixed plane, and from our viewpoint (calculated in Cunitz's *tables for spherical astronomy*), we know where they will appear in the sky. However, that does not tell us where they actually are in their orbital path, only what they look like from our viewpoint on Earth. What the tables of average motions do, in effect, is move our viewpoint from the surface of the Earth, to an imaginary viewpoint somewhere outside of the solar system. This might seem unimportant, but it is an example of a problem that initially had our ancestors believing that the Earth was the centre of the solar system. If you can only view the planets from the surface of the Earth, of course they look as though they are orbiting us. But when you can, in a mathematical sense, stand outside the solar system, you realise that, in fact, the planets, including Earth, orbit the Sun. Knowing the positions of planets in the solar system has important astronomical uses.

An attempt to measure the actual distance of the Earth from the Sun was first made in 1761 using the method of observing the *Transit of Venus*. This involves comparing observations of that planet moving across the face of the Sun, as viewed from different points on the Earth. Clearly, to be able to plan such an observation requires a knowledge of when such a transit will occur. A key part of that is knowing exactly where the planet in question is within its orbit. The 1761 *Transit of Venus* observations were not particularly successful, but a further opportunity in 1769, famously involving Lieutenant James Cook, had better success.

The third and closing section of Maria Cunitz's book consisted of a set of *tables for computation of eclipses*. This section includes a variety of information, primarily for the purpose of calculating the position and time of eclipses.[3] You might wonder why this would be useful? Eclipses are interesting phenomena, but why would we want to be able to predict when and where they occur? Aside from some important uses in astronomy, for example, for investigating long-term variations in the length of the day (currently 365.25 days, but this is increasing by about one-fortieth of a second every thousand years), there are some clever and very practical uses in unexpected areas. One of these is the ability to accurately date events in the ancient world. Many

---

[3]Solar eclipses occur when the Moon passes between the Earth and the Sun, wholly or partially blocking our view of the Sun. Lunar eclipses occur when the Earth passes between the Sun and the Moon, causing the Earth's shadow to obscure the Moon.

solar and lunar eclipses have occurred over the centuries, and many were recorded by our ancient forebears. Using Cunitz's tables, or a modern equivalent, we can look back in time and match historical accounts of eclipses to an extremely precise date. In this way, it has been possible to date the reigns of monarchs and the exact timing of some significant historical events in ancient Greece, Rome, Assyria, and China: all in the pre-Christian era, and as far back as 1000 BCE. For example, the date of the death of Xerxes I[4] has been set at some time between August 4 and 8, 465 BCE thanks to this method.

7.1.2   Why Were Astronomical Tables Invented?

The underlying essence of Maria Cunitz's innovation was a question of *how to locate celestial objects*. Each of the three main sections is concerned with predicting the positions of stars, planets, and their moons from different points of view. The applications of the solution to this problem ranged from the abstract and scientific, to the very practical. Coupled with other key inventions, accurate timekeeping, for example, Maria Cunitz's suite of astronomical tables, played a critical role in bridging the gap between science and its practical application.

7.1.3   How Creative Were Astronomical Tables?

Relevance and Effectiveness: There seems little doubt that Maria Cunitz's astronomical tables were strongly effective. They embodied not only correctness, in that they epitomised the application of the mathematical and scientific knowledge of the day, but they also achieved what they were designed to achieve and did so to a high level of satisfaction. They may have been less accessible to untrained people, but to those knowledgeable in the proper techniques, they were highly relevant and effective. Cunitz's astronomical tables score 4 out of 4 in this category.

Novelty: Here, we must consider how the innovation impacts on our understanding of the problem, and how it may, or may not, shed new light on the problem. It seems clear that these tables were an incremental improvement on Kepler's *Rudolphine Tables*. However, this is not a weakness, and in fact, Cunitz's version set out to overcome some well-known problems with Kepler's tables. This innovation, in other words, was as much about how the knowledge contained in the tables was communicated, as it was about the tables themselves. Part of their value was in showing that a novel approach could increase the accessibility and the utility of the astronomical tables. For these reasons, Maria Cunitz's tables score 4 out of 4 for novelty.

---

[4]Xerxes I, or Xerxes the Great, of Persia was the antagonist in the Battle of Thermopylae, opposed by Macedon's King Leonidas of Sparta. This battle was the focus of the 2006 film 300.

Elegance: This category considers not just the quality of execution of the innovation, which must have been extremely high, but also the presentation, and the way that the information, in this case, fits together. To be useful and useable, these tables had to be clear, understandable, and of course, correctly calculated. They also had to be available in a language that made it easier for a wide audience to use them. This is why Cunitz published her *Urania Propitia* not only in Latin but also in German. She wanted the widest possible audience to use the information. All of these qualities suggest an innovation high in elegance, and I give this example a score of 4 out of 4.

Genesis: The only possible criticism of Maria Cunitz's innovation is that it might appear more incremental than radical. It was a significant improvement over its predecessor, but not so much a radical departure. It is strong in this category, for example, in *foundationality* (it establishes a novel basis for further work) and *vision* (in establishing a new benchmark for judging the calculation of astronomical tables), but does it reach a level in this category that merits a maximum score? In fact, when we consider its wider impact, I think it does. Did Maria Cunitz imagine that her calculations for eclipses would make it possible to date the death of Xerxes I? Did she envisage that her tables for spherical astronomy would underpin the economic and military rise of countries like Holland and England? It is the broader effects, often impossible to envisage ahead of the invention, that show how some innovations solve more than just the immediate problem. I give the astronomical Tables 4 out of 4 for genesis.

Total: We arrive at a total score of 16 out of 16 for Maria Cunitz's astronomical tables! This *very high* rating for creativity also places the innovation at the top of our catalogue, exceeding other inventions like the Buckle and the Propeller. Maria Cunitz's innovation may be a little different from many that we consider: it is less tangible than many, but our analysis shows that it was exceptionally creative.

7.2   The Two-Handed Internal Rotation (1690)

Giving birth is like taking your lower lip and forcing it over your head—Carol Burnett, American Actress (1933–)

The taxonomy for classifying inventions that we have used—energy-, information-, and material-handling systems—works especially well for tangible artefacts. These are the inventions that we can see and touch. Sometimes this categorisation doesn't work as well, or is harder to apply, because the innovations in question are less tangible, or perhaps more abstract. Is a *way of doing* something—a process, in other words—any less a candidate for the label *invention* than a physical object? In fact, we can allocate inventions to one of four types, each of which might handle energy, or information, or material. In addition to the obvious *product*, and the *process* just introduced, we can also describe innovations that are *services*—usually intangible combinations of things like labour and physical aids—as well as *systems*.

In the case of the latter, the term refers to complex combinations of hardware, people, and software.

The two-handed internal rotation is clearly not a tangible artefact. Nor is it a system. Is it then, like pizza delivery or babysitting, a service? Possibly. Or is it a material-handling (in this case, an unborn infant) process? I think it fits clearly into this category, and while there is no artefact per se, there is no doubt that this innovative process was a vital step forward in the practice of obstetrics, in an era that was still beset by myth and mystery.

7.2.1 What Was Invented?

When you consider just what can go wrong during pregnancy and childbirth, I sometimes marvel at the fact that the human species has been so successful in reproducing and growing. However, even with all the knowledge and skill of modern medicine, things can, and do, continue to go wrong. Imagine the situation several hundred years ago, in an era before medical science understood the mechanism of infections, and at a time when doctors shunned obstetrics because it was considered messy and unseemly! Even in the mid-nineteenth century doctors contributed to the transmission of puerperal fever[5] by not washing their hands. Childbirth was dangerous, both for the mother and for the baby.

If infection and other risks are not bad enough, so-called *malpresentations*—circumstances where the foetus is not in the correct position at the time of birth—may account for around 5% of births.[6] Unlike tearing, haemorrhaging, infection, and other problems that may affect either the mother or the baby, malpresentations are a shared risk. If a mother is unable to deliver a baby due to a malpresentation, the consequences can be dire. Today, of course, we benefit greatly from the ability of doctors to perform caesarean sections, but in 1690, this was not a realistic option.

How did expectant mothers cope with all of the dangers and risks of pregnancy and childbirth, for much of human history? The midwife[7] was, indeed still is, a constant in the history of childbirth. In prehistoric times, this may have been simply another woman who had been through the process herself and who could provide some sort of support and help to an expectant mother. Even in ancient Egypt, and

---

[5] Ignaz Semmelweis, a Hungarian doctor, discovered that the incidence of puerperal fever—resulting from uterine infection following childbirth, and responsible for a fatality rate of typically 20–25% of mothers—could be drastically reduced simply by getting doctors to wash their hands! Even when he discovered this link, he had the problem that this was considered insulting to the doctors, and his findings were ignored until some years after his death.

[6] In Australia, in 2010, figures published by the Federal Government's Department of Health stated that 3.9% of *presentations* were breech, 0.2% were face or brow, while 0.7% were shoulder/transverse or compound.

[7] The term midwife is thought to derive from the Middle English (i.e. English from the period 1066–1500) word "mid" meaning "with", and "wife" meaning "woman". A midwife, therefore, was someone who was "with the woman/mother".

certainly Classical Rome and Greece, midwifery had a status and recognition that we can regard as a profession. Midwifery and obstetrics (i.e. the related discipline practised by doctors) would come into conflict in the eighteenth century, as doctors sought to assert their superiority as *scientific* medical professionals. However, in the seventeenth century, midwifery was, in its professional sense, a well-developed discipline, with a considerable body of knowledge accumulated over a long period.

Justine Siegemund (1636–1705) is a prime example of the profession in the latter half of the seventeenth century. Her story is fascinating, as much for the prejudices, jealous accusations and barriers she faced from male doctors, as it is for her publication of *The Court Midwife*, in 1690. In this text, based on her extensive experience (she had first practised midwifery in 1659, delivered more than 6000 babies by her death in 1705, and had worked as the city midwife in Lignitz, and the court midwife to the Elector of Brandenburg). Siegemund set out a rigorous and evidence-based account of numerous problems, and solutions, associated with childbirth. She described, for example, common problems with umbilical cords, and the condition *placenta previa* (where the placenta implants at the bottom of the uterus, potentially blocking the birth canal). However, it is Justine Siegemund's treatment of shoulder presentations, with the *two-handed internal rotation*, that we will focus on as her *invention*.

Shoulder presentations, in Siegemund's time, were frequently fatal to the infant, and potentially so to the mother. Without going into unnecessary detail, any abnormal presentation may make delivery not just difficult, but impossible by normal means. In the days before surgical options, and without any other solution, this was a bad situation, to say the least.

The two-handed internal rotation was simply a way to rotate the baby while it was still inside the uterus, correcting the presentation and allowing a successful birth to take place. Figure 7.2, taken from *The Court Midwife*, makes the method abundantly clear. A sling or cord was attached to one of the baby's limbs (in Fig. 7.2, the leg) and pulled, while the midwife's other hand was used to push the baby's upper body in the opposite direction, thus rotating the infant into a more favourable presentation (a *breech* presentation—not the natural ideal, but better than the shoulder presentation). All of this, of course, had to be done by touch and without any modern diagnostic (e.g. imaging) tools. The procedure had to be done before the uterus began contractions, otherwise the rotation could be difficult, if not impossible to achieve.[8]

7.2.2 Why Was the Two-Handed Internal Rotation Invented?

It is tempting to go with something quite general, such as *how to deliver babies*. However, this is too general, and the solution, the two-handed internal rotation, is

---

[8]Gedoppelter Handgriff der Justine Siegemundin, erstmals publiziert 1690. Kupferstich aus der 2. Auflage ihres Lehrbuches von 1723 (The two-handed internal rotation of Justine Siegemund, first published in 1690. Copper engraving from the second edition of her textbook from 1723). Image Credit: Wikimedia, H.-P. Haack, Creative Commons 3.0: https://creativecommons.org/licenses/by-sa/3.0/legalcode.

Fig. 7.2 Two-handed Internal Rotation

not a solution for some 99% of deliveries. It is the solution to a much more specific problem of delivery. This highlights the importance of defining the real problem, and not a related problem, or a symptom of the real problem. The issue here is more specifically one of *how to correct shoulder presentations*. That is really very specific, but without that level of specificity, in this case, we would risk a solution that misses the mark.

7.2.3   How Creative Was the Two-Handed Internal Rotation?

Relevance and Effectiveness: Siegemund's invention neatly captures the application of our forebears' knowledge of physiology and the mechanisms of childbirth. Midwives in the late seventeenth century clearly knew what was possible, in terms of the manipulation of a foetus in utero, and in the late stages of a pregnancy. What is slightly less clear is the success rate of this method. Did it work in every case of a shoulder presentation that was detected by a midwife, or could complications arise? Was it sometimes difficult, or even impossible to rotate the infant by this means? Did the method risk any injury to either mother or baby? These latter considerations, which address the question of the non-functional performance of a solution (i.e. *how well* it works), must inject some doubt about this innovation. For these reasons, I will give this invention a score of 3.5 out of 4 in this category.

Novelty: The two-handed internal rotation is an interesting case for analysis. Unlike some of our innovations, which show fairly even scoring patterns across subcategories, this one is rather up and down. Novelty is a case in point. It is strong in *diagnosis*—the method clearly highlights the weaknesses of the alternative (doing nothing)—and is also highly prescriptive, showing what can be done to improve on doing nothing. On the other hand, it is notably less strong in how it offers ideas for different applications of the innovation. As we noted in the definition of the problem, it must be quite specific, making the solution similarly less generalisable. Overall, in the category novelty, these peaks and troughs balance out, still leaving us with a strong 3.5 out of 4.

Elegance: In this category, my overwhelming impression is that the solution leans somewhat away from what we think of as *elegant*! What I mean is that it seems hard to think of this solution as pleasing, neat or graceful. It certainly had to be well-executed, which is an important consideration for elegance, but in other respects must have had an element of "it isn't pretty, but it works". I cannot imagine anyone describing the two-handed internal rotation as beautiful, but if it worked, and saved a baby's life, then perhaps elegance is relatively unimportant. In contrast, there is a simplicity and convincingness to the solution. The foetus is in the wrong position, well get in there and turn it. It is almost too simple! Where I end up for this category is the conclusion that elegance is most definitely not tied to appearances. This innovation was uncomplicated and obvious. Anyone can understand the solution, but the trick, of course, was having the experience, patience and skill to do it successfully. I give this innovation 3 out of 4 for elegance.

Genesis: As we saw with novelty, the two-handed internal rotation has some strengths and some more modest qualities for genesis. The method was a novel basis for further work, whether better ways of achieving the rotation or methods (e.g. the caesarean section) that eliminate the need for a two-handed internal rotation. On the other hand, it is less obvious that this innovation is the basis for solving unrelated problems. It was more than just an incremental improvement, given that the *method* it replaced was to let nature take its course! On the other hand, it seems hard to make a case that this invention is entirely radical in nature. It did, no doubt, contribute to a more general theme of some of our medical inventions: namely that our ancestors were not entirely subject to the whims of nature. However, it was probably a relatively small step on that path. I have scored this innovation a 3 out of 4 for genesis.

Total: Justine Siegemund's two-handed internal rotation finishes with a score of 13 out of 16 for creativity, placing it in the *very high* range. This is a good example of an invention that has some room for improvement across all four categories. To be fair, it may have been more effective than I give it credit for, and this may have been further enhanced with a little more elegance. For example, it is possible that *better execution* might have included a cleaner, more hygienic cord, or even a way to rotate the foetus into an even better position for birth. Similarly, both novelty and genesis would be higher if this method had more applications beyond this particular kind of malpresentation. That takes nothing away from the fact that it must have saved many lives and contributed a little more to demystifying birth and medicine in an age that was still beset by uncertainty.

7.3 The Corn Mill (1715)

I will be ruthless in cutting out waste, streamlining structures and improving efficiency—
Theresa May, United Kingdom Prime Minister 2016–2019 (1956–)

Our final invention in the Age of Exploration sits just at the cusp of the next epoch, the Age of Enlightenment, and is nicely placed for the transition into the First Industrial Revolution. Much has been written about this event, and it has been characterised as a social, a political, and of course, a technological revolution. Most people associate this period with the introduction of steam power as the decisive factor in catalysing the transformation; however, the First Industrial Revolution[9] was as much a phenomenon of water power, as it was of steam.

What really happened was not the invention of water or steam power *per se*, but the invention of ways to *exploit* these sources of power. The innovation that we consider now is an example of this exploitation. The corn mill took a process that had been performed in a variety of less efficient and productive ways for centuries—by hand, for example—and scaled it up. Although superficially simply an incremental improvement to an existing solution, this material-handling system had to address a variety of other problems embedded within the job of grinding corn to produce flour.

7.3.1 What Was Invented?

Sybilla Righton Masters lived in colonial New Jersey, her family having moved there from Bermuda in 1687 when she was still a child. At this stage of history, the European settlement of the New World was still in its early days. Jamestown, Virginia, established in 1607, though not the first European settlement, gives us a sense of the development of the colonies at the time of her arrival. This was still a society and an economy largely based around low levels of technology: small-scale, labour-intensive farms, fur-trapping and trading, fishing, forestry, and so on. The unsavoury and immoral use of slave labour, well-established by the late seventeenth century, only reinforced the low-tech nature of the economy.

Changes, however, were afoot. The first stirrings of the Industrial Revolution in Europe (which is regarded as beginning in around 1760) were emerging with the development, in 1696, of a steam pump by Englishman Thomas Savery (see *Homo Problematis Solvendis* for a description). Water, used in simple ways for centuries (water wheels, for example, moved water for crop irrigation, or to grind grain), was available as a source of power. What began to change in this era, however, was the way that these sources of power were utilised. Sybilla Righton Masters' invention

---

[9]For the sake of clarity, the First Industrial Revolution, associated with the introduction and exploitation of steam and water power, ran from about 1760–1830. The Second Industrial Revolution, associated with mass production, assembly lines and electricity occurred between the late nineteenth and early twentieth centuries. The third, associated with automation and computers, began in the 1950s, and the fourth, associated with digitalisation, AI and the like, is happening now!

was an early example of greater levels of sophistication applied first to water power and later to steam power.

Masters invented a new method for processing *cornmeal*. Prior to her invention, maize (corn) was dried and ground into a powder to make cornmeal. The traditional *millstone* grinding method used a stationary, circular, and horizontal *bed stone*, and a moving *runner stone* positioned on top of this, with grain or maize fed into this arrangement through an *eye* in the centre of the runner stone. The runner stone could be turned by a person, an animal, or by water or wind-power. If ground to a sufficiently fine level, cornmeal is sometimes referred to as corn flour, though even in this fine state, it is typically coarser than wheat flour. The resulting cornmeal was a staple food, used for making corn bread, batter, and in colonial times, a type of porridge known as *samp*.

Masters' invention set out to achieve a similar outcome—the production of corn-meal—but through a rather different approach. Instead of the circular grinding action of millstones, her method, also water-powered like many mills, used a cylinder rather like a crankshaft[10] (or camshaft) to move heavy *pestles* up and down into *mortars* containing maize (Fig. 7.3). In this way, Sybilla Righton Masters' corn mill *pounded* the maize into cornflour, rather than grinding it. The resulting corn flour became known as *Tuscarora Rice*, and was a rather coarse-grained flour, used for what is now commonly known as *grits* (a kind of porridge).[11]

Sybilla Righton Masters' story is also significant in the history of *Femina Problematis Solvendis* because she lived in a time and place that did not permit women to own property. This included intellectual property, meaning that she could not, in her own right, apply for or be granted a patent for her invention. Fortunately, her husband, Thomas supported her inventive endeavours, and they filed a patent in his name in London. In the documentation supporting the application, Thomas made it clear that this was, in fact, her invention. In granting the patent, in 1715, King George I also publicly acknowledged her role as the true inventor.

7.3.2   Why Was the Corn Mill Invented?

My first reaction to the task of defining the problem solved by Masters' corn mill is to say this it was intended to solve the problem *how to grind corn*. That definition illustrates the danger of verbs and even nouns that are too specific. Grinding corn places limitations on idea generation that may fail to uncover genuinely creative ideas. Much as *how to screw screws* suggests only a screwdriver as solution, *how to grind corn* makes it hard to think of anything but millstones. The real problem, of course, is broader. Grinding is not the only method, and corn may not be the

---

[10]I discuss the invention of the crankshaft in the twelfth century in *Homo Problematis Solvendis*.

[11]Copy of the corn mill patent issued to Sybilla Masters in 1715. Image Credit: Public Domain. Source: https://wams.nyhistory.org/early-encounters/english-colonies/patent-for-cleaning-and-curing-corn/.

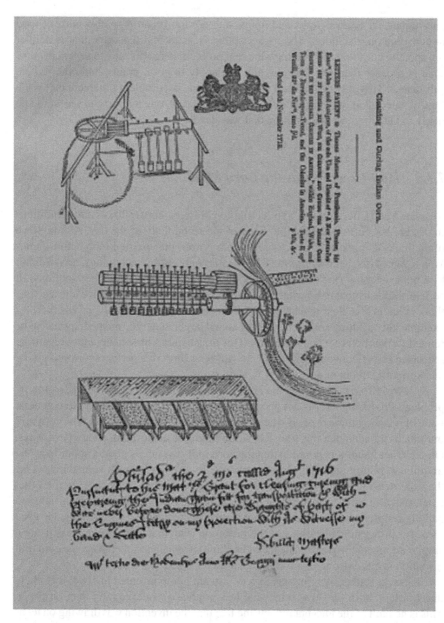

Fig. 7.3 Corn Mill Patent, 1715

only target. Of course, verbs and nouns that are too broad may also make it hard to converge on a solution. The trick, as always, is to find that zone that is neither too specific nor too broad. Perhaps the real problem here was *how to produce flour*? That is broader than just grinding and can apply to corn, grains, and other edible substances. This may not be ideal either, but we need a problem statement that is the logical antecedent of the invention we are considering, one that we can see would be likely to lead to the actual solution that was developed.

7.3.3   How Creative was the Corn Mill?

Relevance and Effectiveness: Sybilla Righton Masters' corn mill was quite ambitious in design, but entirely within the scope of the technology of the day. Although we often think of crankshafts as a tool of steam power, they existed long before James Watt invented his steam engine in 1776. Masters applied this technology, driven either by a horse or water (the details depicted in Fig. 7.3 appear to show both options). It certainly appears that it did what it was supposed to do, pounding maize into a coarse flour. What is less clear is whether it did this uniformly and evenly, or whether, for example, the resulting corn flour still contained larger material, and perhaps had to be refined further before it could be used. This injects just a little uncertainty regarding relevance and effectiveness, so that I am stopping short of a perfect score and giving the corn mill 3.5 out of 4.

Novelty: Broadly speaking, the corn mill helps us understand the underpinning problem. Producing flour is not necessarily a matter of grinding, but a more general issue of reducing the size of corn kernels or grains to a smaller powder. The corn mill introduces the idea that this transformation might be achieved in a variety of ways. On the other hand, it is possible that the corn mill was not as good a solution as the traditional milling approach. In other words, it may have shown something worse than the existing method (a kind of *reverse-incremental* approach). On the other hand, the corn mill does shed new light on possibilities. Greater mechanisation, a paradigm shift from grinding to another form of reducing the maize to flour, and even ideas for different applications of the invention (one that comes to mind is using a similar arrangement to automate metal working). On balance, therefore, the corn mill invention is quite strong in novelty and deserves a score of 4 out of 4.

Elegance: Rather like Thomas Savery's steam pump invented in 1696, it is likely that Masters' corn mill was somewhat less impressive in execution than on paper. This is no fault of the idea but reflects the fact that the invention was drawing on technologies that, although they existed, were still at the leading edge of what technicians of the days were able to produce. Gears, crankshafts, connecting rods, bearings, and the like were still rather new, and depended for their efficiency on good execution. Without similarly well-executed tools, this period sometimes saw gaps between idea and outcome. Inventions in this era were often just a little ahead of their time. Not enough to prevent their execution, but just enough to make the result sometimes a little unreliable. For these reasons, and with no real evidence to the contrary, I feel I have no choice but to give the corn mill 2.5 out of 4 for elegance.

Genesis: In contrast to elegance, Sybilla Righton Masters' corn mill innovation is a strong and impressive example of stepping out of an existing paradigm. The shift away from a very old *rotational*, grinding paradigm to a novel, reciprocating, pounding paradigm not only opened up new possibilities for the production of corn flour but also served as an example of how an external power source, coupled with more sophisticated use of emerging mechanical technologies, could lead to new ideas for automating and scaling up many different traditional processes. Regardless of whether it worked well or not, this invention prised open the door of industrialisation. It peeked through to the other side and saw what might be possible! For genesis, the corn mill innovation scores 4.

Total: The water-powered, reciprocating corn mill invented by Sybilla Righton Masters ends up with a total of 14 out of 16 for creativity. This places it in the *very high* range, reflecting the important role that it played in opening a new approach, and demonstrating the potential of the underpinning technology. Although it would be replaced by methods that were better still, and never fully displaced the traditional millstone method of flour production, this innovation is significant as a herald of the Industrial Revolution and probably deserves more credit in this respect.

Chapter 8
The Age of Enlightenment (1685–1815): Commercialising Innovation

The Age of Enlightenment is a natural sequel to the Age of Exploration. Our problem-solving ancestors began, in the Renaissance, to understand the world around them in a more modern, scientific sense. As this understanding developed and grew, they began to explore more widely, solving problems that were not only physiological and basic, but increasingly driven by more sophisticated needs. Change continued unabated, and many of the upheavals of this era were associated with *reason*, and a general shift in interest away from systems premised on subservience and ignorance, to systems premised on individual liberty and religious tolerance. A general attitude developed, in the Age of Enlightenment, that individuals should dare to know,[1] and the more our ancestral mothers dared to know, the greater the needs that emerged, and the more problems they had to solve. Not quite so enlightened, however, was the attitude to women that still prevailed for much of this era. It was only in the late 1700s, and the early 1800s, that we begin to see women emerging, and acknowledged, as inventors, able not only to *own* their inventions, but to profit from them as well.

Our first innovation in this era is *silk design*. By no means merely an artistic hobby, silk design was a foundation of eighteenth-century European fashion. In this era, we begin to see, very clearly for the first time, the intersection between creativity, innovation, and the *exploitation* of novel ideas for commercial purposes. The second invention we examine in this epoch is the development of *Coade stone*. This was not the first artificial stone used by modern humans, but it had important properties and characteristics that satisfied special needs in the eighteenth century. Finally, we cross the Atlantic to the New World and study another important commercial innovation, in this case as much a response to political changes as any other. *Straw weaving*, like silk design and Coade stone, was not simply a hobby, but an important commercial enterprise, ripe for innovation, but also vulnerable to external forces like politics and conflict.

---

[1] First attributed to Roman poet Horace, in his *First Book of Letters* (20 BCE), the phrase was also used by German philosopher Immanuel Kant (1724–1804) in his essay *Answering the Question: What is Enlightenment?* (1784) and became a catchcry of the Enlightenment period.

© Springer Nature Singapore Pte Ltd. 2020
D. H. Cropley, *Femina Problematis Solvendis—Problem solving Woman*,
https://doi.org/10.1007/978-981-15-3967-1_8

8.1 Silk Design (1740s)

When women are empowered in the design and innovation process, the likelihood of success in the marketplace improves by 144%! – Indra Nooyi, Indian-American Business Executive, Former CEO of PepsiCo (1955–)

The first of our Enlightenment inventions might seem to be a matter of artistic and aesthetic pursuits. Of course, there is no doubt that the silk designs we are considering are not merely functional. There was an undeniable element of satisfying the highest-order needs—self-actualisation and esteem—with this invention, and yet, the solution was more than merely painting pictures of flowers on fabric. Anna Maria Garthwaite was a *designer*, and as such, she is an early example of the intersection between function and aesthetics. We can consider this as an example of the more general nexus between, for example, engineers and industrial designers in the twenty-first century. The former is often said to work *from the inside out*, while the latter work *from the outside in*. Both are concerned with the realisation of solutions, but the engineer's primary concern is the internal functioning, while the industrial designer's principal concern is how it looks and feels, and its usability and the like. What Garthwaite's silk design represents is a shift from a more *function-centric* approach to technological problem solving to a more balanced relationship between form and function. To put this another way, when our early ancestors were concerned only with basic physiological needs—warmth, for example—they had little time to worry about whether their animal skin cloaks looked stylish. As we learned to satisfy these basic needs, it seemed inevitable that our attention would eventually turn to aesthetic considerations as well.

8.1.1 What Was Invented?

In analysing this innovation, we must be careful to separate the material-handling aspects of woven silk design—more a question of *how* the design is produced and what I will call the information-handling aspects. Anna Maria Garthwaite brought a new *style* to the images woven into the silk fabrics as *brocades* (in which the ornamental design is a non-structural weft,[2] giving the impression that the design is embroidered onto the fabric) and *damasks* (in which the design involves typically satin *warp* threads and sateen *weft* yarn).

Anna Maria Garthwaite lived in Spitalfields, the centre of English silk weaving in the east of London, from 1728. Over the following three decades, she created over 1000 designs for silk fabrics (Fig. 8.1 is, of course, a black and white image of the original coloured design). These designs, as was typical of silk designers in this era,

---

[2]In weaving *warp* threads or yarns run the length of the fabric, while *weft* threads/yarns are inserted under and over the warp. Structural wefts hold the fabric together, while non-structural wefts are ornamental.

Fig. 8.1 Design of
Meandering Floral Vines,
attributed to Anna Maria
Garthwaite, c1740

were rendered as watercolour paintings. These were then used by weavers to produce
the actual fabric. A large majority of Garthwaite's designs survive and can be seen
in the Victoria and Albert Museum in London.[3]

What makes Anna Maria Garthwaite's work important for our discussion is that
she did not simply paint flowers as inspiration for silk weavers. The fashion at the time
had been dominated by French interpretations of the Rococo preference for floral
silk designs. These designs used the development of the *points rentrés* technique
in weaving (which allowed a more three-dimensional representation of patterns on
the silk) to create stylised and somewhat unrealistically coloured images of flowers
on the fabric. Garthwaite adapted the points rentrés technique, and the prevailing

---

fashion, to create an *English* style of clustered, more naturalistic flowers, typically spread across a pale background.

From the designs she rendered in watercolour, we learn two important things. Not only did she create a unique style of silk design (the information-handling part of her innovation), but she must also have understood the material-handling aspects, because her watercolours also typically included instructions to the weavers. This is borne out by her adaptation of the points rentrés technique of weaving. In simple terms, Garthwaite invented both a visual style, and the means to render that style in woven silk. We will take these as a package, accepting that the two things work together. What I cannot tell you is which came first: was she a weaver-turned-artist?

8.1.2   Why Was Silk Design Invented?

Whether we take Garthwaite's invention as an inseparable combination of style and method, or treat each separately, we face an unusual situation in our catalogue of inventions. Generally speaking, the innovations that we have considered have always had not only a clear underpinning problem, but also a *functional* emphasis. I want that to mean a *pragmatic* quality, which seems to lean away from aesthetics, but there is a dilemma. If we are using Mawlow's Hierarchy of Needs to help us understand the drivers behind invention, then we cannot pick and choose. Inventions cannot be limited to the lower-order physiological needs simply because those seem to lead to *practical* inventions. Self-fulfilment is just as valid a driver of invention and creativity as the physiological driver for air to breathe. What I am trying to say is that creating an aesthetically pleasing style and technique for weaving floral patterns into silk is just as much a solution to a problem as is making shoes to keep your feet warm. Each can be assessed in terms of how well it solves its problem, how unusual or surprising it is, how well-executed it is, and whether or not it fundamentally changes how we understand the problem. If Garthwaite's driving problem was *how to render designs*, then that is good enough for our purposes.

8.1.3   How Creative Was Silk Design?

Relevance and Effectiveness: Anna Maria Garthwaite's silk design package was highly effective. Both the style she created, and the supporting adaptation of method, were feasible and successful. Alongside the *technical* feasibility and effectiveness, we can also consider the *popular* effectiveness of her designs. As I indicated earlier in this case study, higher-order needs were part of the driver for this innovation. People wanted to dress in nice fabrics with stylish patterns that reflected their aesthetic preferences. Garthwaite created a solution that, to judge from its popularity across Europe and North America, was highly competitive with the French designs of the same period. Of course, her solution was aided, to some degree, by economic sanctions that blocked French imports to North America in the era (giving her designs something of an unfair advantage in that market), so that there is a tiny sliver of doubt that accompanies part of our judgement. For this reason, I have to acknowledge a slightly uneven playing field with a score of 3.5 out of 4.

Novelty: Part of our judgement of novelty revolves around the way an invention helps us to understand the other solutions to the same problem. Concepts like *shortcomings in other solutions* may be clear for more tangible, concrete inventions, but seem less helpful in situations where a solution is basically a subjective preference for style. In other words, I feel confident we could all agree that Edison's light bulb draws attention to the shortcomings of candles as a form of lighting. We could objectively state what those shortcomings are. However, I feel less satisfied when I apply this judgement to the notion of a *style*. Not because the criterion is flawed, but because the solution itself is too subjective. This makes it hard to say if Garthwaite's style was novel. The subjectivity of the solution would make it much harder to reach a consensus. I might think it is highly novel, whereas you might disagree. My judgement here is that Anna Maria Garthwaite's stylistic innovation was not especially novel, even if her methodological innovation was. I therefore come out somewhere in the middle of our scale with a 2.5 out of 4 for novelty.

Elegance: In contrast to my uncertainty regarding novelty, I think there is clarity in the case of elegance. Stylistically, this seems clear. You may not particularly like the style, but you can still appreciate the quality of execution, the harmoniousness of the design, and so on. Even the methodological aspects of Garthwaite's invention—the adaption of the weaving technique—must also score well. Although we know less about this aspect, it seems highly likely that she developed a careful, clear, easy-to-understand process that allowed weavers to convert her designs from painting to the finished fabric. I therefore give Garthwaite's innovation 4 out of 4 for elegance.

Genesis: The same concern I had for novelty resurfaces when we consider genesis. In some ways, this category is easier. Anna Maria Garthwaite's invention was largely incremental, both stylistically and methodologically. It does not seem to have made any radical change to the underlying problem: how to render designs (on silk). It also did not set a radical new basis for further work, nor did it offer ideas for solving other, unrelated problems. While it did, in some sense, establish a new norm for judging other similar solutions (i.e. the French style), the assumption of a new, *better* norm may be absent here. It was different, rather than inherently better. These general qualities—an element of *different*, not necessarily *better*—detract somewhat from genesis. This was a rather narrow, incremental (and subjective) innovation, unlike many that we have seen. I therefore give it 2.5 for genesis.

Total: With a total of 12.5 out of 16, Garthwaite's silk design sits just between *high* and *very high* creativity. I think the basic issue for this innovation lies in the greater subjectivity of the solution. This may be a feature of solutions to higher order needs. A solution to the problem of keeping me warm is generally pretty clear. A solution to the problem of making me feel happy is inherently more personal and subjective. We still judge creativity in the same way, but the solution itself may inject a degree of uncertainty that draws away some elements of creativity, especially novelty and genesis. Anna Maria Garthwaite's silk designs were feasible, technically successful and well-executed, but not everyone might have found them beautiful and ground-breaking.

8.2   Coade Stone (1770)

Each of us is carving a stone, erecting a column, or cutting a piece of stained glass in the construction of something much bigger than ourselves – Adrienne Clarkson, Governor-General of Canada (1999–2005) (1939–)

The *idea* of artificial stone was not new in 1770. The Romans, for example, invented a form of concrete—opus caementicium—in the third century BCE. For centuries, humankind has understood the value and utility of artificially produced, stone-like material for creating durable structures. In fact, in the early twenty-first century, twice as much concrete is used, by weight, compared to steel, aluminium, wood, and plastic combined. What is remarkable about our current innovation—Coade stone—is the special qualities and properties of artificial stone that Eleanor Coade created, and the circumstances surrounding her work.

Great Britain in the late eighteenth century was the birthplace of the Industrial Revolution. Although we frequently associate this with advances in manufacturing, thanks to the availability of water and steam power, the broader impact of the Industrial Revolution in Great Britain was social, economic and even political in nature. It was also during this period that Britain was expanding geographically with colonies across the globe. The general picture, therefore, was of a society on the rise. Like other expanding societies before it, Ancient Rome, for example, a very practical, outward sign of growth was *infrastructure*. Indeed, this is the case today. We can get some sense of the state of a country's fortunes just by looking across the skyline of one of its cities. Count how many construction cranes you see. Strong economies require new buildings, new roads, new bridges, and other infrastructure.

Throughout human history, societies that enjoyed wealth, growth, and success have not only built for functionality. They have also celebrated their achievements with grand monuments. Whether temples and tombs, statues and stadiums, or figures and façades, humans have always sought to add aesthetics to effectiveness. These structures, however, could hardly achieve their purpose if they crumbled and decayed. How would future generations know of the greatness of Ancient Egypt, Greece, and Rome if the Sphinx, the Parthenon, or the Colosseum had crumbled to nothing within a few centuries? Monumental architecture, even today, thus requires materials that will stand the test of time. In Great Britain, in the late eighteenth century, there was much to memorialise, and Eleanor Coade had a solution for this purpose.

8.2.1 What Was Invented?

In 1769, Eleanor Coade had been running her own business in London, working as a linen draper,[4] for several years. Her entrepreneurial spirit does not appear to have been dampened by the death of her father in that year, or his second bankruptcy. Indeed, in the same year Eleanor bought a struggling artificial stone factory from one David Pincot, and within two years was firmly in charge of *Coade's Artificial Stone Manufactory*. She fired Pincot in 1771, over some disagreement about his conduct, and over the subsequent five decades the business enjoyed great success.

The artificial stone market in this era was quite crowded. There was strong demand for a material with two key qualities. First, the material had to be durable. Second, these artificial stone materials had to be capable of being moulded. The key driver behind this was the fact that *carving* statues (and other structures like building façades) from durable rock was slow, difficult, and expensive. In an era where there was much to celebrate and memorialise, who could wait for a statue to be hewn from granite?

In Great Britain in the eighteenth century, there were a number of different artificial stone recipes. Most of these were some form of ceramic,[5] but none had the exact properties needed for this application. Eleanor Coade didn't invent the basic concept of artificial stone, but she is believed to have perfected both the composition of the raw materials, and the process of firing (i.e. setting and hardening) the finished structures.

Coade stone (which we can classify as a material-handling system) was, technically speaking, a form of *stoneware*. This is a special type of ceramic, based on clay, fired at high temperatures, and with a non-porous finish. Coade's recipe consisted of: 60–70% so-called *ball* clay (this contains certain proportions of substances such as quartz); 10% crushed soda lime glass (the most common type of glass, used in windows and bottles); 10% *grog* (also called *firesand*, and containing a high proportion of silica and alumina); 5–10% crushed flint, and 5–10% fine quartz. This mixture could then be moulded and shaped as required, before undergoing a four-day firing process in a kiln, at 1100 °C.

Statues and other architectural features made from Coade stone are generally renowned for their durability, often faring better than natural stone from the same period. Eleanor Coade's business enjoyed over forty years of success. For the last eight years of her life, the business was managed by a distant cousin, William Croggan. We have a clear indicator of the success of the business at the time Eleanor Coade died, in 1821, because Croggan bought the manufactory from her estate for £4000 (equivalent to about US$ 500k today). By 1833, the business was in decline, eventually ceasing operations in the early 1840s, as newer, better materials superseded Coade stone.

---

[4]*Drapers* were wholesalers or retailers of cloth, usually used for the purpose of making clothing (as opposed to making sails, for example).

[5]Ceramics include earthenware, porcelain, and brick.

Fig. 8.2 South Bank Lion Statue, London

Even today there are many famous examples of Coade stone in Great Britain. These include the Royal Pavilion in Brighton, parts of Buckingham Palace, and the South Bank Lion statue at the southern end of Westminster Bridge, in London (Fig. 8.2).[6]

8.2.2   Why Was Coade Stone Invented?

This is a tricky one. That may be due, in part, to the fact that Coade stone addresses more than one issue. In fact, the problems that Coade stone solves fit neatly into a phrase that I have frequently used to describe a driver behind incremental innovation. In engineering, we often say that incremental developments are focused on making something *better, faster, and cheaper*. That seems to be exactly what was driving the development of Coade stone. In this case, *better* means more durable than natural stone (and other forms of artificial stone), *faster* means the ability to replace the slow process of hand-carving stone, and cheaper means replacing hand-carving with something that did not have the same high manual labour cost.

To stay consistent with our format of *how to verb noun*, the trick is to find a single verb and noun that combine *better, faster, and cheaper* into one. In this case, I think our problem is *how to replace carving*. Implicit qualifiers, i.e. constraints, include the fact that our focus is the carving of stone (and not wood, for example), and that *replace* implies better, faster, and cheaper.

[6]Image Credit: Wikimedia, Public Domain.

8.2.3   How Creative Was Coade Stone?

Relevance and Effectiveness: It seems hard to find fault with Eleanor Coade's special form of artificial stone, both in the more technical assessment of creativity, and in an everyday sense. Hers was the only successful artificial stone business in the era, suggesting that there must have been something about the effectiveness of her unique recipe that set it apart from others. We also have the, dare I say *concrete*, evidence in the examples of statues and other architectural features that are still in existence, some 200 years after their creation. If we need a more technical assessment, then Coade stone was technically feasible, did what it was supposed to do, and did it well. 4 out of 4 for effectiveness!

Novelty: Coade stone unequivocally scores strongly in problematisation. Imagine pitching this product to a potential client. You can produce statues, for example, faster than a sculptor can carve them, that will last longer than carved stone, and will be cheaper to make. The weaknesses of the traditional method, in other words, are starkly revealed by this new solution. Set against this, however, is that fact that Coade stone is not a radically new approach (*reinitiation*) and does not offer a fundamentally new perspective on solutions (*generation*), precisely because it is so incremental in nature. Combining these elements of novelty together, we still get a very strong 3.5 out of 4 for this category.

Elegance: I am tempted to give Coade stone the maximum in this category. However, we need to avoid being seduced by the qualities of the end product. The invention here is not the statues and ornaments, but the means for producing these. One of the challenges of Coade stone was the rather careful mixing of the basic ingredients, and the slow and careful firing process. In fact, we know that there are some examples of Coade stone structures that did not fare so well over time, possibly because of difficulties in controlling the process. This injects just a little inelegance in this invention. To be really strong in this category Coade stone would be simple to make. The fact that it required a fair degree of care and control detracts a little from the simplicity and harmoniousness of the innovation. I am reflecting that in a slightly imperfect score of 3.5 for elegance.

Genesis: Did Coade stone change how the problem (in this case, what we defined as *how to replace carving*) is understood. I have already highlighted the strongly incremental nature of Coade stone, and that must weaken the case for more radical or disruptive qualities in this category. Coade stone was more a pinnacle in the evolution of artificial stone, rather than a radical new starting point. It took prevailing concepts, recipes, and methods to some degree of perfection and would remain largely unchallenged in this space for 50 years. Another way we can see a relative lack of genesis is in the vulnerability of Coade stone to subsequent innovations. When something better came along, in the mid-1840s, Coade stone really had no answer. There was no further value to be squeezed out of this paradigm, and it was replaced by something better. It was certainly not weak in this category, but, in the long term, it lacked enough genesis to prevent it being replaced. I give it 3 out of 4.

Total: With a total of 14, Coade stone is comfortably in the *very high* range for creativity. It was highly effective, there is no doubt. However, as I have already argued, it was a highly incremental innovation. This enabled it to dominate the problem space for a long period, but in the end, left it vulnerable to something even better, or even completely different. If it had been a little easier to manufacture, this would add half a point, but a higher score was not really open to something that was very good at solving a very particular problem.

8.3   Straw Weaving (1809)

You see, when weaving a blanket, an Indian woman leaves a flaw in the weaving of that blanket to let the soul out – Martha Graham, American Dancer and Choreographer (1894–1991)

We close out the Age of Enlightenment with another material-handling system that is a perfect example of change driving innovation. Like many innovations, however, it was not a technological change, but a social and political change that provided the stimulus. The early nineteenth century was a time of global upheaval. Europe was mired in conflict, with France and her allies, under the leadership of Napoleon, competing with Great Britain and her supporters for geographic and economic control as far afield as the South Pacific. The USA, only recently independent, was attempting to remain neutral in this conflict. In 1805, Great Britain, however, blockaded French ports in an attempt to starve the French economy of vital imports. For the USA, this created a problem. France was one of the most important trading partners of the USA, and the British blockade meant not only that American merchant ships were prevented from delivering their goods to France, but also that these ships, and their cargoes, might be seized by the British.

In 1807, US President Thomas Jefferson responded to these economic sanctions by placing his own ban on the import of British goods into the USA. The effect, unfortunately, was to cripple the US economy. Unable to trade effectively in continental Europe, and now without a trade relationship with Great Britain, the US economy crashed, with vital income-generating exports—cotton and tobacco, for example—dropping by about 80% in the space of a year.

Imports to the USA were also affected. Unable to purchase many manufactured goods from Europe, women in the USA found their access to straw bonnets, among other things, suddenly impacted. The *fashion* of the time saw nearly all women wearing head coverings, for a variety of reasons. Whether to keep their hair tidy, or to protect it from dust, or for reasons of style or religion, straw bonnets were extremely common across Europe and North America in this period (see Fig. 8.3). This meant that there was a significant industry associated with the production and supply of straw bonnets. In Britain alone, the straw plaiting and hat making industries

Fig. 8.3 Straw and Silk
Bonnet, c1810

supported the livelihoods of thousands of people at this time. The USA had to find a way to meet this demand, now that it could no longer rely on imports.[7]

8.3.1 What Was Invented?

In 1798, Betsy Metcalf, at the age of only 12, was frustrated at the cost of the beautiful imported straw bonnets. She decided to pull apart examples and figure out how they were made. In the process she taught herself how to plait, or braid, the straw, how to cut it, how to wet it in order to soften and shape it, and how to knot the straw to hold the structure together. Rather than patent the technology—something that had become a *theoretical* possibility for women in the USA after the passage of the 1790 Patent Act[8]—Betsy chose instead to share her ideas freely. This created the underpinnings of a local straw weaving and hat making industry in the USA.

---

[7] Image Credit: Metropolitan Museum of Art, Costume Institute. Creative Commons CC0 1.0 Universal Public Domain Dedication.

[8] Although inventions could be patented in the US after 1790, in practice, women could not own property at this time. Inventions are *intellectual* property, which meant that until laws began to change, starting in around 1821 (the state of Maine passed a law allowing women to own and manage property in their own name during the incapacity of their spouse) either a female inventor had to get a man to seek the patent in his name, or simply give up!

When the effects of the trade embargo struck, in 1807, the scene was set for local manufacturing to replace the importation of straw bonnets. This is where Mary Dixon Kies comes into the picture. Kies developed a method for weaving silk ribbon or thread into the straw, creating a decorative effect that soon became highly popular. Taking advantage of an inconsistency in the 1790 Patent Act, and its later amendments—which stated that *any person or persons* could seek protection of their ideas—Mary Kies applied for, and was granted, a patent for her invention in May 1809. The *inconsistency* here was with the general property laws of the day. By *not* limiting patents to men—the Patent Act also explicitly stated that applications can refer to ideas that *he, she, or they* might have invented—this statute may have helped spark a change in the rights of women. Mary Kies became the first women in the USA to be granted a patent. Her invention helped to stimulate the local straw weaving and hat making industry, struggling under the trade embargo, with improvements to the cost-effectiveness of the manufacturing process, as well as a boost to popularity thanks to the stylistic improvements. We also cannot ignore the fact that Mary Kies's invention contributed to a change in the rights of women.

8.3.2   Why Was Straw Weaving Invented?

Straw weaving, of course, was not invented in 1809. Our particular focus here is on a new kind of straw weaving, incorporating improvements in both method and style. Mary Kies contributed one step in a long line of incremental developments to this industry, prompted by changes that were both economic and social in nature. It is easy to be impressed by *radical* innovations. The first time one of our ancient forebears showed her family the wheel, it is fair to say that their minds would have been blown. It is also easy to forget the less spectacular, but no less important, incremental improvements that frequently are vital for turning a radical (but imperfect) idea into a *practical*, cost-effective, impactful solution to a problem.

So it is with this current innovation. Mary Kies didn't invent the idea of weaving straw into hats. That was a long-standing concept. It had been invented, improved, and turned into a practical, productive industry for many decades. What Kies did do was, responding to a unique set of circumstances, at a unique point in time, solve a particular problem. *How to revive weaving*, for a local, straw bonnet industry struggling with cost-effectiveness issues, trade restrictions, and difficulties of demand, makes Mary Kies not just an inventor, but an entrepreneur as well. She solved not just the technical problem, but the wider economic and social problem, throwing in some improved rights for women, for good measure.

8.3.3 How Creative Was Straw Weaving?

Relevance and Effectiveness: There is a strong similarity between the current invention and the previous one concerned with silk design. Both inventions combine a functional, methodological component with an aesthetic component. Both, in fact, were influenced in some way by the economic issues associated with the long-standing conflict between Great Britain and France. Mary Kies's invention stimulated the straw weaving industry in North America just when this was needed. It was effective both as an innovation in straw weaving and hat making, and as a means of reviving the industry. In all respects, this invention was highly effective and scores 4 out of 4 in this category.

Novelty: Once again, rather like silk design, there are some questions in relation to novelty. In one sense, Mary Kies's invention was very incremental in nature, with novelty that must be regarded as modest (in a constellation of very creative inventions). An improved design, more attractive and easier to make, is certainly not uncreative. However, it is no radical breakthrough, and indeed, when fashion tastes changed again in the early 1800s, Mary Kies would ultimately die penniless, as the demand for bonnets based on her innovation declined. In another sense, however, we see glimpses of an innovation that was about more than just the bonnets themselves. At least some of Mary Kies's innovation might be tied to the way that businesses, and whole industries, respond to change. Mary Kies therefore deserves some credit for an innovation in how industries respond to change. Nevertheless, I feel that the primary conclusion here is of rather limited novelty that was, itself, vulnerable to further change, and I give her invention 2.5 for this category.

Elegance: The innovation in straw weaving and hat making introduced by Mary Kies is strongly elegant. Like silk design, the consideration here is both aesthetic and functional. The innovation must have been pleasing, graceful, and harmonious in a visual sense, in order to stimulate the desire and demand for locally made straw bonnets. Equally, the *process* of creating these bonnets, with their woven-in silk, must have been complete, well-executed, and skilfully developed in order to achieve the cost-effectiveness and ease of manufacture that made it possible for the fledgling industry in the USA to adopt these changes. This gives us a clear score of 4 out of 4 for elegance.

Genesis: Unlike silk design, Mary Kies and her straw weaving innovation has some more far-reaching qualities in the category of genesis. The innovation itself was rather incremental, but the impact of the innovation was deeper, more diverse, and more long-lasting than that of silk design. If genesis is a matter of how an invention changes our understanding of the problem—perhaps causing us to ask, "what is the *real* problem here?"—then we see this in a couple of ways in the straw weaving example. Not in terms of the actual weaving problem, but in terms of the wider issues surrounding this. Straw weaving is not just a matter of turning up and plaiting some straw to make a hat. To do this successfully requires a market for straw hats. That market might be affected by things as diverse as a conflict in a far-off country. So perhaps the real problem is not really anything to do with straw weaving directly, but to do with politics? Another way that Mary Kies and her innovation changed how we understand the problem was the matter of property ownership, women's rights, and the granting of patents for inventions. Perhaps the real problem that Mary Kies solved was not about straw weaving at all, but the question of how to improve the rights of women? I have given this innovation 3.5 out of 4 in this category to recognise the more far-reaching consequences of Mary Kies and her desire to improve the manufacturing of straw bonnets.

Total: We know little about Mary Kies's invention. The details of her patent were lost in a fire at the US Patent Office in 1836. As we have discussed, however, the true impact of her invention may have been less technical and more social. Despite only scant technical details, the broader impact of the innovation leaves us with a *very high* score of 14 out of 16 for creativity.

Chapter 9
The Romantic Period (1800–1900): Industrial Innovation

The Romantic Period—running roughly from the late 1700s to the late 1800s (or approximately, the nineteenth century in its entirety)—is probably characterised best by the consequences of industrialisation. This era has also been referred to as *The Great Divergence*,[1] and while there is debate about exactly when it began, there is little disagreement on the consequences. This was a time of accelerating and major change, facilitated by ever-increasing technological prowess, and is therefore most evident in the so-called Western world (i.e. Western Europe and North America). Electricity, as a usable utility, emerged in this period, along with many of the resulting innovations that are fundamental to modern life: sanitation systems, refrigeration, railways, telegraphic communications, telephones, automobiles, and of course the electric lightbulb. In many respects, the Romantic Period saw the fruits of the Industrial Revolution emerge, though not without many social, economic, and environmental costs.

In *Homo Problematis Solvendis*, I began this era with a major innovation in transportation, the *velocipede*.[2] This came about as a result of a lack of food for horses. In the present volume, we will begin this era with another horse, but not the kind you might expect! Although it will seem an odd invention to start an era characterised by industrial innovations, the *Berkeley Horse* reminds us that our Maslowian needs are not confined to problems of transport, communication, and the like. Psychological and self-fulfilment needs have always been with us, sometimes overwhelmed by more basic needs, but always present. The Berkeley Horse is an invention that caters to some of our more *intimate* needs.

The second invention in this period is critical to much of what controls our lives, especially at the dawn of Industry 4.0. Ada Lovelace's *computer program* set the stage for how we translate instructions in natural, everyday language, into the algorithms that drive our modern computers. The next invention in this era is the *propeller*. This innovation played a critical role in the development of transportation, commerce,

---

[1] See the book *The Great Divergence* by Kenneth Pomeranz (2000).
[2] Very simply, the first bicycle.

© Springer Nature Singapore Pte Ltd. 2020
D. H. Cropley, *Femina Problematis Solvendis—Problem solving Woman*,
https://doi.org/10.1007/978-981-15-3967-1_9

and industry. It is also a story of a woman breaking down traditional stereotypes and setting an example for her peers.

Fourth on our list of inventions in the Romantic Period—you get a bonus invention in this era—is another sea-faring innovation: the *life raft*. This invention is an excellent example of the interactions between needs, problems, and solutions. As our ancestors began to master activities like sea travel, bringing this under ever-greater control with steamships, propellers, and sophisticated navigation, it is probably inevitable that attention would turn from merely travelling by sea, to travelling *safely* by sea. The life raft is wholly a creation stemming from the invention of ships. One solution creates new problems, and these new problems give rise to their own solutions, and so the process continues.

9.1   The Berkeley Horse (1828)

The more sex becomes a non-issue in people's lives, the happier they are—Shirley MacLaine, American Actress and Author (1934–)

Readers may have noticed that there are *four* inventions in the Romantic Period and not the usual three. The reason for this bonus is simple. After selecting my thirty *Femina* inventions, a female colleague sent me details of this one. Although the subject matter is a little bawdier than things like artificial stone, wheels, or computer programs, there are some important ideas and lessons embedded in the Berkeley Horse. One lesson is what this device says about the human needs that drive invention and problem-solving. The Berkeley Horse does not fulfil basic physiological needs, such as the need for air or water. Rather, it operates at the higher level of psychological, and perhaps even self-actualisation, needs. It is a different kind of invention, and a different kind of problem, but these are no less powerful as drivers of creativity and innovation. I cannot quite make up my mind as to the classification of this invention. Is it a system that handles information or material, or is it, in fact, an energy-handling system? For the reason that it converts the chemical energy stored in muscles into a force applied to the body of the *victim*—however the resulting pain is interpreted—I think it has to be classified as an energy-handling system!

9.1.1   What Was Invented?

The Berkeley Horse was devised by Theresa Berkeley, a well-known dominatrix in nineteenth-century London, in 1828. Berkeley catered to members of the aristocracy, both men and women, and specialised in what is now generally referred to under the banner of BDSM.[3] Her special skills lay in the practice of inflicting pain for pleasure, and her discretion, coupled with the exclusive nature of her clientele, ensured that she became very wealthy. In fact, at her death in 1836, her estate was estimated to be £100,000 (over US$13 million in today's money)!

The device itself (as shown in Fig. 9.1), also known as a *chevalet* (after the French for *easel*), consisted of a padded board, on which a person could be restrained, typically face down. The angle of the board could be opened or closed, rather like a ladder, to suit the activity, and included openings at strategic points to allow access to various parts of the anatomy. I think you get the picture![4]

Fig. 9.1 Berkeley Horse, c1828

The Berkeley Horse

9.1.2 Why Was the Berkeley Horse Invented?

Understanding *what* was invented, in this case, does not require a lengthy discussion—the contemporary drawing in Fig. 9.1, inclusive of birches, tells us everything we need to know. What is more interesting, and informative, is *why* it was invented.

In many ways, the story of the Berkeley Horse is a microcosm of the *process* of innovation. Generally speaking, *change* is a driver that accompanies factors in Maslow's hierarchy of needs. Assuming that the underpinning need is psychological (I refer you back to Fig. 1.1), for example, a need for love, friendship, intimacy, and even perhaps some sort of pain, then why not simply beat a client with a birch? Why the need for a special device like the Berkeley Horse? This is where change enters the discussion. Perhaps the change was social: changing attitudes to sex, for example, that opened up a new demand for certain forms of gratification? The point here is universal to the innovation process. Underpinning needs, in the sense of Maslow, are satisfied in a variety of ways, at various points in time. Changes (social, demographic, economic) occur creating new versions of the underpinning need. Once upon a time, sexual gratification (as an expression of a psychological need) was satisfied, for example, by extramarital sex. Change occurred, resulting in a refinement of the need, and a new problem demanded a new solution. Instead of *how to achieve sexual gratification* in a *vanilla* sense, the problem for some people evolved into *how to achieve sexual gratification through bondage, domination, and sadism.*

Theresa Berkeley understood this change and the new problem that it created. Like any good inventor, she understood that it needed a new solution. Furthermore, she understood the commercial value to be gained in finding a new solution to this problem.

9.1.3 How Creative Was the Berkeley Horse?

Relevance and Effectiveness: It would be difficult to argue for anything less than a score of 4 out of 4 for this criterion. If Theresa Berkeley's business success is any indicator, then it must have met with considerable customer approval. We know, from contemporary writings, that her clients were prepared to pay her £1 (about US$130 now) "for the first blood drawn"; £2 (US$260) "if the blood runs down to my heels"; £3 (US$390) "if my heels are bathed in blood"; £4 (US$520) "if the blood reaches the floor"; and £5 (US$650) "if you succeed in making me lose consciousness". To think that flogging was a common and feared punishment in this era! Theresa Berkeley, furthermore, would employ female assistants to help her. One, for example, might be seated inside the device in order to provide other forms of pain and/or pleasure to Berkeley's clients.

Novelty: While the Berkeley Horse has some very strong elements in this category—in particular, the way it draws attention to weaknesses in other, existing, solutions—it loses some points when we assess other elements of novelty. This device makes it obvious that either a table, or a vertical board, lacks the flexibility and options offered by the Berkeley Horse. On the other hand, the easel concept is perhaps only a relatively simple incremental improvement to these other options. Thinking back to the idea of points of diminishing returns, we could say that the Berkeley Horse advances further up and along a curve, not yet reaching the point where further improvement is redundant, but it is not far off. My point is that the Berkeley Horse is an extension of existing concepts but is *not* a radically original approach. It is incrementally novel, but not radically novel, and so I cannot give it more than a good 3 in this category.

Elegance: Our task of judging the quality of execution of the Berkeley Horse is slightly hampered by the fact that we are relying on contemporary drawings of the device. It is unclear if any examples exist today, and I am unable to find any photographs. Having said that, it is hard to imagine that Theresa Berkeley would have enjoyed the success she did, or maintained her elite clientele, if the device had been poorly executed. In other words, and notwithstanding the fact that the purpose was to help her inflict pain on her clients, it must have been a high-quality piece of work. We can assume that the padding on which the customer leaned was comfortable and probably made from good-quality leather. Similarly, the structure itself must been well made, robust, and sturdy. It would not have done her reputation any good if the horse collapsed and injured someone mid-thrashing! In fact, I assume the device was made of the best materials, polished, and finished to a standard appropriate to the social standing of her clients. In addition, some care and attention must have gone towards ensuring that the various holes—feet, face, and other vital parts of the anatomy—were correctly positioned. All of this suggests a well-proportioned and well-executed device that added to the functionality. For these reasons, it gets a 4 for elegance.

Genesis: For this criterion, we return to considerations of newness, and how the solution impacts on our understanding of the underlying problem. Against all of the elements of genesis, the Berkeley Horse is reasonably strong, but not perfect. It contains elements that might give us ideas for further, new refinements. It gives us a fresh perspective on the basic problem: the client has to neither stand up nor lie down. It gives us a new standard for judging devices of this type. Where it is not quite, so strong is that it probably fails to offer ideas for solving unrelated problems! Nevertheless, it performs quite well in terms of genesis, and I have given it a score of 3.

Total: With a total score of 14 out of 16, the Berkeley Horse is comfortably inside the *very high* range of creativity. The commercial success that Theresa Berkeley enjoyed seems to attest to the strong relevance and effectiveness of the invention, and this was reinforced by a strong degree of elegance—it must have been well executed to achieve the robust functionality that it enjoyed. Where it loses a little in terms of creativity is the combination of novelty and genesis. One weakness regarding the latter is *transferability*. I find it hard to make a convincing argument that the Berkeley Horse offered ideas for solving unrelated problems. It was very much a one-trick pony (forgive the pun)! In terms of novelty, its minor weakness is that it is really an incremental solution. The idea of strapping someone to a frame or table was not new. What was an improvement in this innovation was achieving that restrained in a way that offered some flexibility for the related activities of BDSM. In other words, an adjustable easel concept is an improvement to the existing concept, rather than a paradigm-breaking, radical notion.

9.2 The Computer Program (1840s)

Never trust a computer you can't throw out a window—Steve Wozniak, American Engineer (1950–)

It can be tempting, through the lens of our twenty-first-century perspective, to regard artefacts, services, processes, and ideas from a bygone era as inherently inferior to those that we now produce. This may be particularly so in the case of something like a computer program, given the strong modern, digital connotations of this technology. The solid-state, semiconductor transistor was only invented in 1947. Programming languages such as FORTRAN and BASIC only appeared in 1957 and 1964, respectively. Digital computers, in the sense of the desktop machines we now think of, only came into being in 1981.[5] How is it possible that a *computer program* could have existed in the 1840s?

Perhaps the easy way to understand this is to remind ourselves that a computer program is, in its simplest form, really only a set of instructions, in this case for a machine. This is no difference from a set of instructions for a cook—a recipe for baking a cake, for example—or a set of instructions for assembling a flat-pack bed. The only real difference is that the computer, in these cases, is a person. The point is that humans have been describing complex tasks and processes—writing instructions—for centuries. The pioneering computer program, invented by Ada Lovelace, is simply an information-handling system designed to give instructions to the analytical engine invented by Charles Babbage, thereby unleashing the full potential of that device.

9.2.1 What Was Invented?

Charles Babbage conceived of his *analytical engine* in the early 1830s. This general-purpose, programmable, mechanical, steam-driven (literally, not figuratively), digital

---

[5]This date is significant for creativity and innovation because it also provides a neat illustration of the impact of new inventions. We usually focus on the benefits, but innovation also frequently destroys things that have gone before. The introduction of the PC in 1981 quickly led to the introduction of digital word processing. This, in turn, destroyed the market for typewriters. The company Smith Corona had been an innovative typewriter manufacturer since 1886 but was more or less wiped out by the digital word processor.

computing machine was revolutionary in concept. It consisted of four main elements: a *mill* (analogous to a modern CPU), a *store* (analogous to modern computer memory), a *reader* (the engine's input device), and a *printer* (for generating output information), all recognisable as components of modern, digital electronic computers. Although not completed by Babbage, largely owing to its complexity (if we were scoring it here, it would fare quite poorly under the category *effectiveness*, precisely because it was not entirely feasible at the time of its development), it set in motion important changes in our ancestors' understanding of *computing* (so it would have scored well for *novelty* and *genesis*). Our interest, of course, is not really the analytical engine at all, but one key part: the reader.

Ada Lovelace—more correctly, Ada King, Countess of Lovelace—is our focus here, and it was her work on getting the instructions *into* the analytical engine, via its reader, that gives us our present innovation. To understand her contribution, which is unaffected by Babbage's failure fully to implement his device, we must digress a little to understand the *reader*. Babbage's input device consisted of punched cards and card-reading technology developed by Joseph Marie Jacquard for his Jacquard Loom. This device was an incremental development in weaving that allowed looms to create intricate fabric patterns such as tapestries, damask, and brocades. Babbage recognised the potential of this card-reading technology for entering instructions into his analytical engine (so he loses some marks for outright novelty). The key for Ada Lovelace, therefore, was the punched card system of entering instructions.

As some readers (of this book) will know from practical experience (punched cards were still in wide use in computing as recently as the 1980s), getting from the sequence of instructions you wish to give the computer to the information on the punched cards is no easy task. Like any computer language ranging from FORTRAN through to Python, programs are generally written in a human-readable, natural language form. Nowadays, these instructions are then interpreted or compiled (i.e. converted into a machine-readable form) at which point the computer can execute (i.e. run) the program. In the days before editors and other programmer-friendly, human–machine interfaces that can spellcheck and debug your code, how did the programmer move from the concept (e.g. I want to create a program to add strings of numbers, compute their averages, and list these in order from lowest to highest) to the actual holes punched in cards? This is where Ada Lovelace bursts onto the scene.

Figure 9.2 shows us an excellent example of Lovelace's invention. The *program* we can see in the figure (which is believed to be her first published program) is written for the computation of Bernoulli numbers.[6] What she created was an *algorithm*, in other words, an explicit, unambiguous set of instructions for a computer, written as a sequence of steps that the computer is able to carry out.[7]

---

[6]Bernoulli numbers are, strictly speaking, a sequence of rational numbers. They are used in a variety of complex mathematical processes, such as the Euler–Maclaurin formula. Why? I am afraid you'll have to do some research or consult a mathematician!

[7]This image is from Sketch of The Analytical Engine Invented by Charles Babbage by Luigi Menabrea with notes by Ada Lovelace, 1842. Image Credit: http://www.sophiararebooks.com/pictures/3544a.jpg, Public Domain.

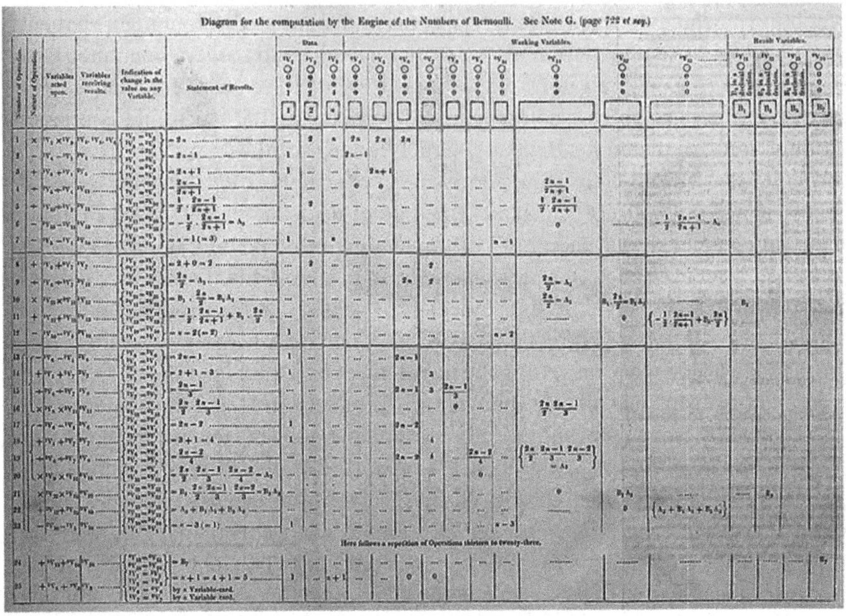

Fig. 9.2 Lovelace's program for the computation of Bernoulli numbers

Unlike natural language, in which we might give an instruction like "go over there, pick up that ball, and bring it back to me", an algorithm generally must operate from a much more restricted dictionary. If the computer only understands instructions like "move", "add", "next", and so on, then your natural language instructions must be converted into this simpler form if a computer is to execute them. Because computers also take our instructions extremely literally (i.e. there is no room for ambiguity), the instructions must leave no room for error. Ada Lovelace's contribution was to show how to translate the instructions needed to perform a complex mathematical calculation into the explicit sequence of steps—the algorithm—that would allow the analytical engine to perform this calculation.

9.2.2   Why Was the Computer Program Invented?

I am not being facetious when I say that the problem tackled by the computer program was *how to program a computer*! We could be a little more general and say that the problem we are solving is how to communicate with a computer or, perhaps, how to translate instructions, but the purpose of the computer program really is very clear. The verb here—program—we know is shorthand for getting instructions into the computer. It is also implicit that these instructions must be in a form that the computer understands. Rather than beat around the bush and say *how to get machine-readable instructions into a computer*, it seems nice and concise to say simply, how to program a computer. It is coincidence, more than anything, that the solution happens to be very similar in name! This would be analogous to soap being called *dirt remover*!

9.2.3 How Creative Was the Computer Program?

Relevance and Effectiveness: Ada Lovelace's computer program presents us with a minor difficulty. This is a difficulty that is also manifest when we talk about Leonardo da Vinci and his *inventions*. Creativity, as we know, demands relevance and effectiveness as a prerequisite. In other words, if an innovation is not effective—that is, if it does not do what it is supposed to do—then it cannot be called creative. Now, there is really only one way to establish effectiveness. You have to make the thing and test it! Leonardo's big problem was (and is) that his *inventions*, as far as we know, were never actually built. So, it is all very well to draw a picture of what looks like a helicopter in the late 1400s, some 450 years before the first real helicopter was built, but that doesn't mean da Vinci *invented* the helicopter. That takes nothing away from the quality of his ideas, but invention is a practical, problem-solving activity. If all that is needed to be a successful inventor is to sketch an idea, then I can now claim to have invented a matter transporter. I have a very nice sketch that I can show you!

Now Ada Lovelace's difficulty here, which takes a small amount of her score for effectiveness, is that her program never actually ran on Babbage's analytical engine. Not through any fault of hers, but because Babbage was never able to finish the thing. Her program was correct and accurately reflected contemporary knowledge, but it loses a little under performance and appropriateness, because, while we think it would have worked, she never got the chance to actually show that it did. Regrettably, this leaves the computer program with a score of 3.5 out of 4 in this category.

Novelty: Unlike effectiveness, the originality of Ada Lovelace's innovation is, happily, unaffected by Babbage's failure to execute his solution. The computer program that she devised is strong in problematisation, in the sense that it helps to define the problem (how to program a computer) more clearly. For example, it shines a light on the deficiencies of other related devices. What I mean here is that the computer program shows how relatively poor were the existing methods of giving instructions to machines. A case in point would be the Jacquard Loom that we have already discussed. The instructions that could be given to that device were extremely limited and inflexible, while the instructions that could be given to the analytical engine, via the computer program, were much more flexible and diverse. Both used the same method of entry, the punched card, but it is the underpinning versatility of the instructions, and the method of encoding them, that sets Lovelace's method apart.

Notwithstanding these strengths, there are some weaknesses in this category with respect to the light that this innovation sheds on the problem at hand. One in particular is that the computer program is probably not a *radically* new approach to encoding a set of instructions for calculating things like Bernoulli numbers. That may sound like blasphemy to a computer scientist, but Ada Lovelace's program was more the pinnacle of expressing an existing mathematical algorithm in a clear set of steps, rather than a radically new approach. She did not invent new mathematical concepts, but simply found an efficient way to encode what was already known. Similarly, it may have been difficult for observers at the time to see other ways that the computer program might be used. Of course, it is used for other types of program and for instructing a computer to do things other than calculate Bernoulli numbers, but those other things remain largely variations on a theme and not radically new ideas. These are relatively minor quibbles, and the computer program still scores strongly for novelty at 3.5 out of 4.

Elegance: The strongest aspect of this strongly creative invention is undoubtedly its elegance. Look closely at Fig. 9.1, and you can see the clarity, care, rigour, and skill that has gone into setting out the precise, step-by-step details of the Bernoulli number algorithm. It is not just a matter of "add this number to that" and so on. Lovelace specifies the variables that are to be acted on, those that hold the result of each step, the nature of each exact step, and so forth. No doubt twenty-first-century programmers could use the detail in Fig. 9.2 to write a modern version of this algorithm with little trouble. It is well executed, complete, and fully worked out, giving it a maximum score of 4 for elegance. This reinforces our view that, had Babbage succeeded in making his analytical engine, this program would have executed successfully.

Genesis: As we return to consideration of newness and the impact of inventions on our understanding of the problems they seek to solve, there is much to admire about Ada Lovelace's computer program. It was highly foundational, in the sense that it provided a fundamentally new basis for further work. We can see this very clearly in the subsequent development of computer programming and the abundance of languages that have emerged. Lovelace's invention is, in some ways, rather like the invention of the cuneiform alphabet that I analysed in *Homo Problematis Solvendis* in this respect. Cuneiform itself disappeared, but it established a new paradigm which has been developed, extended, and exploited ever since.

If this innovation has any weaknesses with respect to genesis, it may be that it is somewhat less transferable than might be the case. While it is true that the general concept can be applied to more than just mathematical algorithms, it remains limited by the basic constraints of the computer itself. In other words, as we know, you can only program a computer to do the things it is capable of doing. Similarly, it is slightly less germinal than it could be, limited by the fact that it does not establish a new way of looking at mathematical problems. It is not a new way of doing mathematics, just a faster and better way. Even with these minor faults, the computer program is a strong example of genesis and scores 3.5 out of 4.

Total: The computer program developed by Ada Lovelace, by which we mean the *methodology* that she developed for entering complex instructions into the analytical engine, scores 14.5 out of 16 on our scale of creativity. There are no obvious areas of weakness, with the lowest individual score being 3.5. Where I have deducted half a point, it has been for factors often beyond the control of our *Femina Problematis Solvendis*. Had Babbage succeeded in making a working model of the analytical engine, and had Lovelace lived longer and benefitted from seeing her innovation in action, it may well have achieved a score of 15.5, if not 16. As always, however, we try to ignore the hindsight that we could apply in these assessments, and try to confine ourselves to the effectiveness, novelty, and so forth that would have been evident at the time the invention in question was introduced. As a result, Ada Lovelace's work sits well inside the range of *very high* creativity and may yet prove to be one of the most creative in our catalogue of female inventions and creativity.

9.3 The Propeller (1869)

We were wavering around like a ship without a sail—Judy Holliday, American Actress, Comedian, and Singer (1921–1965)

The latter phase of the Romantic era—the late 1800s—is an interesting period in the history of technology, in part because it sits astride the Second Industrial Revolution. Readers will recall that the introduction of mechanisation, water power and steam power, from the mid–late 1700s, sparked a surge of innovation. However, the introduction of electricity as a viable source of power for homes and businesses, in the late 1800s,[8] created all sorts of new possibilities and sparked an enormous surge in innovation. It is no surprise that the notion of the assembly line, and the advantages of mass production, leapt forward in this period.

One of the by-products of this Second Industrial Revolution was an acceleration of trade and globalisation. In the late 1800s, that meant ships, and the faster ships could move goods around the world, the more profitable they were. Clipper ships, first introduced in the mid-nineteenth century, and powered only by sail, were one solution to this problem. However, as steam power continued to develop, and as wooden-hulled ships gave way to larger iron-hulled vessels, the next logical step was a method of propulsion that was independent of the vagaries of wind. Ships powered by steam would be unconstrained by the whim of winds. They would be able to sail faster and more directly to their destinations. They could also better exploit new developments like the Suez Canal, opened in 1869, further cutting the journey times for global trade. However, steam power is one thing, but turning this into an efficient way of pushing a ship through the sea is another. Paddle wheels, in fact, were first used for this purpose, but there was a better solution that would prove to be much more efficient for this purpose. That brings us to the *propeller* and the work of Henrietta Vansittart.

---

[8]The first centralised power station, in a modern sense, was opened by Thomas Edison in New York, in 1882.

9.3.1 What Was Invented?

Henrietta Vansittart (1833–1883) was the daughter of an English engineer and inventor, James Lowe, who worked, rather unsuccessfully, in the field of ship propulsion. In 1838, Lowe patented the design for a novel type of screw propeller, but legal complications surrounding the design thwarted his efforts, and by the time Henrietta was in her late teens, the family ran into financial difficulties. Despite these difficulties, Lowe, now assisted by Henrietta, continued to work on the propeller with some success. Although she was not formally trained as a scientist or engineer, Henrietta became increasingly involved in the work, for example accompanying her father on board the vessel HMS Bullfinch during trials of a refinement of the propeller in 1857.

Lowe's propeller was in wide use in naval vessels in the late 1850s, but tragically he was killed in an accident in London in 1866. Rather than abandoning the work, Henrietta took over responsibility for her father's work and began working seriously on the project from this point. Within two years of his death, Henrietta was granted a patent for a new design—the Lowe-Vansittart propeller—not only improving her father's design, but finally overcoming the legal dispute that had prevented her father from benefitting financially from his design.

The Lowe-Vansittart propeller (Fig. 9.3) was not merely a simple arrangement of flat blades spinning in water. Although such an arrangement might work for something simple such as a ceiling fan, in water, and in attempting to propel a ship, there are many other considerations that affect efficiency and speed. One problem for propellers in water is that they may *churn* the water. Imagine the propeller spinning vigorously, but merely frothing up the water, without generating any forward motion. Critical to the propeller's efficiency is that it must move smoothly through the water, with as much of the energy of the propeller being devoted to pushing the water at right angles to the direction of the spin. The problem here is analogous to the design of aircraft wings, which must be shaped in such a way as to generate lift, and slip efficiently through the air, minimising drag.[9]

One of Henrietta Vansittart's contributions was, in her own words, "to economise the power required in driving the steam-propellers for ships or other vessels". She goes on to explain, in the notes accompanying her patent, that "This is effected by so modifying the form of the blades of screw-propellers as to cause them to act more effectually on the water, and to prevent them from "churning" or uselessly stirring the water near the centre of motion".

Vansittart's design proved itself in competition with other designs and soon came into wide use in steam-powered ships of the Royal Navy, as well as merchant vessels. A Lowe-Vansittart propeller was fitted to the famous *Lusitania*, sunk by a German torpedo in May 1915.

Henrietta Vansittart's role in the development of the propeller attracted wide attention, in part because, as a married woman in an era where few women of her education and social standing worked, let alone left the home, she stood out as independent,

---

[9]This image is from the British Patent (#2877) issued to Henrietta Vansittart in 1869. Image Credit: Public Domain.

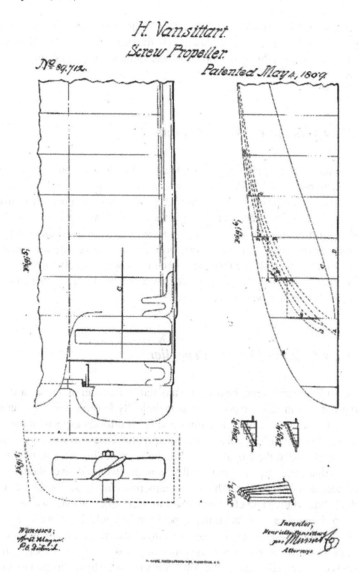

Fig. 9.3 Henrietta Vansittart's Screw Propeller design

entrepreneurial, and inventive. The attention she received does not seem, however, to
have been negative (despite her involvement in a reasonably high-profile affair with
novelist and parliamentarian Edward Bulwer-Lytton). Her engineering achievements
were reported in *The Times* and received many awards at international exhibitions
and scientific meetings in London, Dublin, Paris, Sydney, and other cities throughout
the 1870s and into the early 1880s.

Henrietta Vansittart did not renew the patent for her invention, an obvious energy-handling system, when it became due not long before her death, in 1883. Nevertheless, the design by this self-educated innovator was influential, successful, and widely acclaimed.

9.3.2   Why Was the Propeller Invented?

The Lowe-Vansittart propeller was invented for the clear purpose of solving the problem *how to propel ships*. Prior to the invention of propellers and paddle wheels, the main form of ship propulsion had been the sail. Oars, of course, had also been used in some vessels since ancient times. Sails and oars, however, both suffered from a dependency on unreliable sources of power. Whether the unpredictability of the wind or the limitations of human rowers, both systems were ripe for replacement by a much more powerful and reliable source of power: namely steam. Henrietta Vansittart's innovation took the propeller to new levels of efficiency that ensured sail had no choice but to give way to steam.

9.3.3   How Creative Was the Propeller?

Relevance and Effectiveness: Henrietta Vansittart's innovation scores a clear and deserved 4 out of 4 in this category. It was technically feasible and drew on at least rudimentary knowledge of fluid dynamics and other relevant aspects of the physics that underpins this form of propulsion. In more practical terms, it was widely adopted by a major stakeholder, the Royal Navy, who had compelling reasons only to adopt technologies that were effective and reliable, and that would deliver them an advantage over potential adversaries. This is perhaps particularly so for Great Britain's navy, which had enjoyed unchallenged dominance since its victory at Trafalgar in 1805. No doubt aware of its strategic position, it could ill-afford to rest on its laurels, especially given the technological changes that were underway. The Lowe-Vansittart propeller improved speed and efficiency in steam-powered ships, meeting the demands of (or, equally, fitting the constraints imposed by) an important end-user and stakeholder.

Novelty: The Lowe-Vansittart propeller is very close to a perfect score across the four categories of creativity. The fact that it is so close makes it easier to say what is slightly lacking, rather than extolling its virtues. I have given this invention 3.5 out of 4 for novelty. The primary reason I could not give it a score of 4 is that it is, like a number of other high-scoring innovations, more incremental than radical. It is important to remind ourselves that this is not, in itself, a terrible condemnation of an invention. A good invention—one that is creative—must do its job, and it must be new. The point here is that there is new, and then there is *new*. Incrementally new, namely finding new ways to extend the known, making existing solutions better, or faster, or cheaper, is good and laudable (and valuable). Radically new, however, just goes one step further. The minor weakness of incrementally new is that incrementation has a finite limit. You reach a point where it becomes impossible (or uneconomical) to increment any further. When that point is reached, the only option is to change to a new paradigm and innovate radically. Thus, radical innovation will always have the last laugh and must therefore be slightly better in the long run. The Lowe-Vansittart propeller, in its day, was a sizeable incremental improvement over its predecessors. It opened up a new and better understanding of the physics of propulsion. However, it did not fundamentally change that paradigm.

Elegance: Henrietta Vansittart's invention must have been an extremely carefully and well-executed design. Indeed, one of her chief competitors was rejected by the Royal Navy due to vibration problems. That might seem an innocuous weakness, but remember that to achieve no, or minimal, vibrations in a large, heavy piece of rotating metal requires not only a precise design, but also a very precise execution. Vansittart's design, with carefully and deliberately shaped blades, was no simple matter to construct. In addition to these qualities, it also looks balanced and rounded. It is well-proportioned and aesthetically pleasing. It seems to prove, as so many other inventions do, not only the correlation between effectiveness and elegance, but also the simple dictum that good solutions usually look like good solutions, 4 out of 4 in this category.

Genesis: As was the case for novelty, the Lowe-Vansittart propeller is very close to a perfect score. It sets a foundation for further work, expanding our understanding of the factors that matter in propeller design. It may well have influenced unrelated fields like flight, still very much in its infancy, by developing early principles of how objects behave in fluids (whether water or air). It also helps us to understand and quantify what was weak in previous propeller design and, indeed, other preceding methods of propulsion. However, although it did many of these things, my assessment is that it did not quite do them to the point where the minds of observers, seeing this for the first time in 1869, would have been blown! The propeller paradigm was already established. This innovation took it forward, but did not break it, and for that reason it scores a very strong (but not perfect) 3.5 out of 4 for genesis.

Total: I am trying hard to find reasons to tip this innovation up a notch. It worked well, and it was well designed and well executed. It had many qualities that would be the springboard for further significant improvements, and it shifted propeller design more firmly onto a scientific footing. Had it been a wholly new approach in its own right, then we might be looking at the second maximum score in our catalogue of *Femina Problematis Solvendis*. As it is, the Lowe-Vansittart propeller scores a *very high* 15 out of 16 for creativity.

9.4   The Life Raft (1880)

...O hear us when we cry to Thee, For those in peril on the sea—William Whiting, English Writer and Hymnist (1825–1878)

The final invention that we consider in the Romantic Period is one that was driven by the same underlying forces as the propeller. Electricity, mass production, and the other features of this period drove industrialisation and globalisation. Intra-continental trade—within Europe, for example—could be done by rail, but international trade, in the days before aircraft, could only be done by sea. In fact, even today, some 90% of the world's trade is by sea, for the simple reason that it is still the most efficient and effective way to move large volumes of raw materials and manufactured good long distances. You only need to think of oil, wheat, iron ore, and coal: a single oil tanker can carry about 270 million kg of oil. A single 747 cargo plane can carry about 140,000 kg of goods. The ship, in other words, carries as much as more than 1900 jumbo jets.[10]

Even today, with our modern ships, and advanced navigational systems, sea travel can still be dangerous. In the late 1800s, however, it was far riskier, and when something goes wrong on a ship, the final resort is to take to the lifeboats. We therefore turn our attention to Maria Beasley and her invention of the *life raft*.

9.4.1   What Was Invented?

Maria Beasley was a nineteenth-century American inventor and entrepreneur. Her prolific inventive activities stand in sharp contrast to the work of Henrietta Vansittart (and her propeller) that we encountered in the previous section. The contrast is not in the nature or quality of their work, but in the different environments that surrounded their activities. Henrietta Vansittart lived and worked in Victorian England, where, as we have seen, her inventive activities were seen as something remarkable. Married women from good families don't go around running businesses and inventing things! Maria Beasley, on the other hand, does not seem to have attracted the same sense of amazement. We can attribute this to a rather different social climate that existed in the USA in the same period. This is not to say that women enjoyed radically different treatments in these two countries (*neither country* allowed women to vote, for example, until well into the twentieth century), but the general climate in the USA seems to have been more open and supportive of innovative and entrepreneurial women.

---

[10]Nowadays, of course, we are also more aware of the relative levels of pollution, especially carbon emissions, associated with different forms of travel. Considering only the carbon emissions per kilometre travelled, ships emit about one-third as much CO_2 as planes. We could perform some more detailed calculations to see the relative emissions per kilogram of something like wheat or iron ore, to show why it makes sense to transport these bulky commodities by sea.

Maria Beasley first invented a barrel-hooping machine (designed to improve the speed of manufacture of barrels), for which she was granted a series of patents, in 1878. She evidently made a significant amount of money from this invention and was active in a range of other areas, from cooking pans to a device designed to stop trains derailing. However, it was her efforts to improve the life raft that resulted in a patent in 1880, for a design that sought to improve the fire-proofing, size, safety, and ease of launch of existing life raft concepts. Before Maria Beasley applied her inventive skills to this problem, life rafts typically consisted only of flat, wooden panels, with little in the way of features designed to maximise the survivability of passengers.

The Beasley design, seen in Fig. 9.4, introduced collapsible guard rails; metal floats, seats, fittings for oars; a more streamlined hull; and compartments for food and water. In one variant, she even created guard railings on both sides of the raft, so that if the raft was overturned, it could still be used. The metal floats were hinged so that the base could be folded and made more compact for storage.[11]

What is interesting about this design is that it demonstrates an interesting blend of an existing general concept (a floating raft as a safety device for ships), pre-existing supplementary elements (e.g. storage for food and water, oars, seats), along with two wholly new components. The latter are described by Maria Beasley herself in the patent document and are: (a) the hinged metal floats that form the base of the raft and (b) the reversible guard rail that means the raft can be used with either of its two flat surfaces upright.

9.4.2 Why Was the Life Raft Invented?

The general concept, of course, was already in existence when Maria Beasley turned her attention to its improvement. The general problem was *how to save lives*, with the overarching constraint (that typically accompanies any problem statement, answering the supplementary question; *how well?*) probably focused on saving those lives *at sea*.

9.4.3 How Creative Was the Life Raft?

Relevance and Effectiveness: The crux of relevance and effectiveness is whether or not the invention in question did what it was supposed to do. Did Maria Beasley's improved life raft save lives at sea? Did it do so better than previous designs? Unlike some of our previous innovations, we might have one very good and objective measure of this for the Beasley life raft—a real case of a sinking ship. No speculation and no educated guesswork needed for a documented case. The famous RMS Titanic

---

[11]Image Credit: http://shells-tales-sails.blogspot.com/2015/04/l-is-for-life-raft-inventions-by-women-z.html, Public Domain.

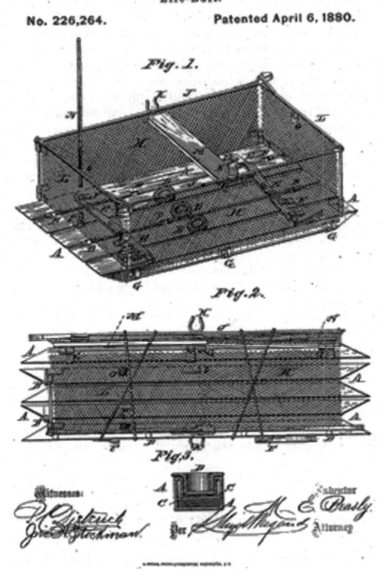

Fig. 9.4 Maria Beasley's life raft patent, April 1880

has been quoted as such an example.[12] Unfortunately, this seems to be inaccurate or rather overblown. In fact, if the four collapsible lifeboats carried on the Titanic were versions of the Beasley design, then something went badly wrong. Only two were launched shortly before the vessels sank. This may have been just part of the general chaos that reigned on the Titanic and which affected the launching of all the ship's lifeboats, but it may have had something to do with the design. Were they, in fact, hard to launch or regarded as inferior to the other lifeboats on the ship? This doesn't fill me with confidence, and I am inclined to reflect this in the score. The Beasley design gets a 3 for effectiveness.

Novelty: The Beasley life raft has some interesting qualities in this category. It is strong in the elements that I call *problematisation* (i.e. the solution helps to define the problem). In other words, the design makes the weaknesses of previous designs very clear. On the other hand, the Beasley design is not quite as impressive in the elements of propulsion (i.e. the extent to which it sheds new light on the problem). Here, the issue is that the life raft *does not* take a radically new approach and *does not* offer any insights into new and different ways of using the artefact. Like many of our inventions, it is largely incremental in nature. Therefore, it scores a healthy, but not outstanding 3 out of 4 for novelty.

Elegance: In many ways, Maria Beasley's life raft is a well-executed innovation. In a narrow sense, it is well made and complete and has a degree of gracefulness to it. Then I ask myself what would it be like to abandon ship and spend hours or even days on a Beasley life raft, in the freezing cold North Atlantic. My answer is: not pleasant! I have a slight unfair advantage here, having served in the Royal Navy and having spent some time in modern life rafts as part of my training. My criticism, which must have been evident to some people even in 1880, is that the design provides little protection from the elements for the people on board. There is no doubt that it addressed some problems with previous designs, but it failed to tackle others. For this reason, I regard it as somewhat incomplete, and unconvincing, and therefore, lacking some elegance. I am giving it a 3 out of 4 for this category.

Genesis: Perhaps not surprisingly, the Beasley life raft scores a 3 for genesis as well. This reflects similar minor, but important, deficiencies in the other categories. Broadly speaking, it goes some way towards changing how the problem is understood. It draws attention, for example, to the size and ease of use of the life raft (and not just the fact that it is present or not). It also encourages us to consider the possible advantages of new materials that might make the life raft better in some way. These

---

[12]If you search the Internet, you may come across stories that claim Maria Beasley's life rafts were used on the Titanic. The Titanic carried only 20 life*boats*. These comprised two wooden cutters (wooden row boats capable of seating 40), 14 standard wooden lifeboats (larger than the cutter, seating 65 people), and four collapsible canvas lifeboats (that seated 47 people). Photographs of the latter exist, and they do not appear to be the same as the life rafts shown in Fig. 9.3. They may, however, be later models based on the original Beasley design. What seems clear is that the Titanic *did not* carry 20 of Beasley's life rafts, as some stories claim, and therefore *did not* save 706 lives. If the four collapsible boats were, in fact, later Beasley models, then her innovation did save some lives (although only two of the four were launched successfully). Unfortunately, therefore, the Titanic story doesn't give us a convenient, objective measure of the Beasley design's effectiveness.

can be seen as the first steps that have led us to some very interesting modern life raft and lifeboat concepts today.

Total: Maria Beasley's life raft scores 12 out of 16 for creativity. This places it at the top of the *high* range, but highlights some weaknesses, evenly spread across the four categories, that hold it just outside the *very high* range. It appears that it could have been designed somewhat better, perhaps drawing on what we now call the *user experience*. What evidence may exist for its real use is somewhat unclear. Maybe it didn't perform quite as expected. Therefore, better execution, and a better focus on the key elements of the problem, may have bumped its score up a couple of points. Beyond that, there was the opportunity to begin a shift in the paradigm. Was the future of life rafts/lifeboats a matter of improving the existing design, or was there the possibility of a radically new approach. Was this more than just a folding, double-sided version of the old style, or would it depart from this concept?

Chapter 10
The Modern Age (1880–1950): Bras, Bombs, and Bits

The Modern Age, defined largely as the first half of the twentieth century, saw some of the best, and the worst, consequences of humankind's inventive ability. Few families even today are untouched, in some way, by the effects of the two great, global conflicts of this epoch, and few were spared the economic effects of the Great Depression. At the same time, from these catastrophes emerged a world poised to reap the greatest benefits of creativity—massive improvements in health, enormous improvements to education, trade and welfare, and the promise of an end to global conflicts with the invention of a mechanism for resolving disagreement and conflict on a global scale. However, an interesting question that often arises is: Would the innovation of this period have been the same without World Wars 1 and 2? There is some compelling evidence that wars stimulate technological creativity, perhaps due to a greater appetite for risk and fewer organisational barriers. Were the advances of the Modern Age accelerated because of these conflicts, or would they have proceeded just the same? Of course, we will never know, but it does draw attention to some of the *non-cognitive* factors that impact on creativity and innovation. How does the environment you live and work in impact on your ability to come up with new ideas and to invent?

Although an era dominated by conflict, the first invention we consider could not be gentler. The *brassiere* reminds us that the first half of the twentieth century was also a time of growing liberation for women. One very simple example of this was that women in the early 1900s began to question some of the cultural norms that told them what they should wear. As fashions changed, women like Mary Jacobs decided that uncomfortable corsets were no longer meeting their needs for practical undergarments, and she invented a very simple, practical alternative. The second innovation in this era is, in contrast, entirely motivated by conflict. However, it is an ingenious example of the cross-fertilisation of radically different fields. What do torpedoes and pianolas have in common? A lot, if you are trying to solve the problem of how to prevent the enemy jamming your controlling radio signals.

© Springer Nature Singapore Pte Ltd. 2020
D. H. Cropley, *Femina Problematis Solvendis—Problem solving Woman*,
https://doi.org/10.1007/978-981-15-3967-1_10

We end this epoch with another vitally important invention that set the foundation of the present Digital Age. The *compiler*, responsible for allowing us to *talk to* computers in English-like language (rather than in 1's and 0's), arose from the early modern digital computers that emerged just before and after the end of World War 2. Although these machines initially were developed for the war effort in Britain and the USA, it became apparent in the late 1940s that the machines had much more general, commercial applications. Programming them, however, would require the work of Grace Hopper and her compiler.

10.1   The Brassiere (1914)

What do I think will help a lot of people and most certainly will help me?—Marion Donovan, American Inventor (1917–1998)

We hear a lot now about the looming impact of digital transformation. As the world moves into a period in which the potential of artificial intelligence, machine learning, and related technologies is being realised, there are vigorous debates about the impact this is having on the work that humans perform. The phrase *the future of work* represents a discussion of the skills and abilities that humans will need in order to thrive in an environment in which robots, for want of a better phrase, have taken over many of the jobs that humans used to perform.

One aspect of this discussion that seems to attract broad agreement is that creativity will remain the job of humans. That's both good and bad news. It is good, because it clearly identifies a role for people in the future of work—we are good at finding new and effective solutions to problems—but it is also bad, because we aren't necessarily educating children in this vital competency. That aside, the reason creativity is the job of humans is because artificial intelligence cannot do what humans do, which is to notice, for example, that it would be more comfortable when walking around if we had some sort of covering on our feet. Similarly, AI cannot recognise that shoulder presentation births could be avoided with a two-handed internal rotation, or that ash and fat mixed together make it easier to clean things. The brassiere is a similar example that illustrates why creativity is the business of *people* and, in this volume, women! How on earth could an artificial intelligence work out that the lives of women could be improved, and made more comfortable, by fashioning some fabric into a support for their breasts? This creativity comes from humans—women, of course—noticing a problem and devising a novel and effective solution.

10.1.1 What Was Invented?

Mary Jacob (later known as Caresse Crosby), an American perhaps better known for her role as a publisher and for her patronage[1] of writers such as Ernest Hemingway, Anaïs Nin, and Charles Bukowski, invented the backless brassiere in 1910 at the age of only 19. The driving force behind her innovation is delightfully practical and down-to-earth, illustrating, only too well, the relationship between change, new problems, and the development of new solutions to those problems.

In 1910, long before her bohemian and literary exploits in Europe, Mary faced what might seem to a modern reader, as a rather self-indulgent problem. She came from a wealthy family and was in the midst of attending, as was typical at the time, a series of debutante balls.[2] As was the fashion at the time, women wore a stiffened corset and corset cover under their dresses. These were designed to be very tight at the waist, to hold the wearer very upright, and to flatten and push together the breasts. Of course, these were part of the fashion of the time, but were, by all accounts, very uncomfortable for many women (i.e. those whose figures did not welcome being clamped, squashed, and otherwise manipulated into a very restricted shape). Mary's problem, perhaps reflecting the tension between old styles of undergarment and newer styles of dresses, was that she wanted to wear a gown with a plunging neckline that meant the corset and corset cover were visible. Having already worn this arrangement once, Mary was keen to hide her undergarments (not, it seems out of a sense of modesty, but to maximise the visual impact of her gown). An added factor was probably another very pragmatic one. Mary had large breasts and probably found the corset more uncomfortable than many women.

The result of these problems—not having her corset visible under her gown and relieving her discomfort—caused Mary to take action. She took two handkerchiefs and some pink ribbon, and quickly created what we now know as a simple bra (see Fig. 10.1).[3]

Mary's design used loops of ribbon as shoulder straps, and other ribbons passed around the back and tied at the front. Her design quickly attracted the attention of her friends, and after she was offered money for one, she realised the commercial value and filed a patent, receiving recognition and protection of her design from the US Patent and Trademark Office in November of 1914. The design was highly suited to the emerging fashion of plunging necklines and low-cut backs in women's gowns, as well as being suited to a range of different sizes and shapes (of the women wearing the brassiere). Mary's design was not the first brassiere—others having been patented from the 1860s—and was not even particularly successful (she sold the business she

---

[1]Crosby was a central figure in the so-called *Lost Generation* of mainly American writers living in Paris in the 1920s. The story of her exploits, living with her husband Harry on the proceeds of his substantial trust fund, and involving drugs, wild parties, an open marriage, and even a suicide pact is an interesting counterpoint to her inventive abilities!

[2]These events were where young women who had reached adulthood signalled their eligibility to young men in their social circle.

[3]Image Credit: http://www.uspto.gov, Public Domain.

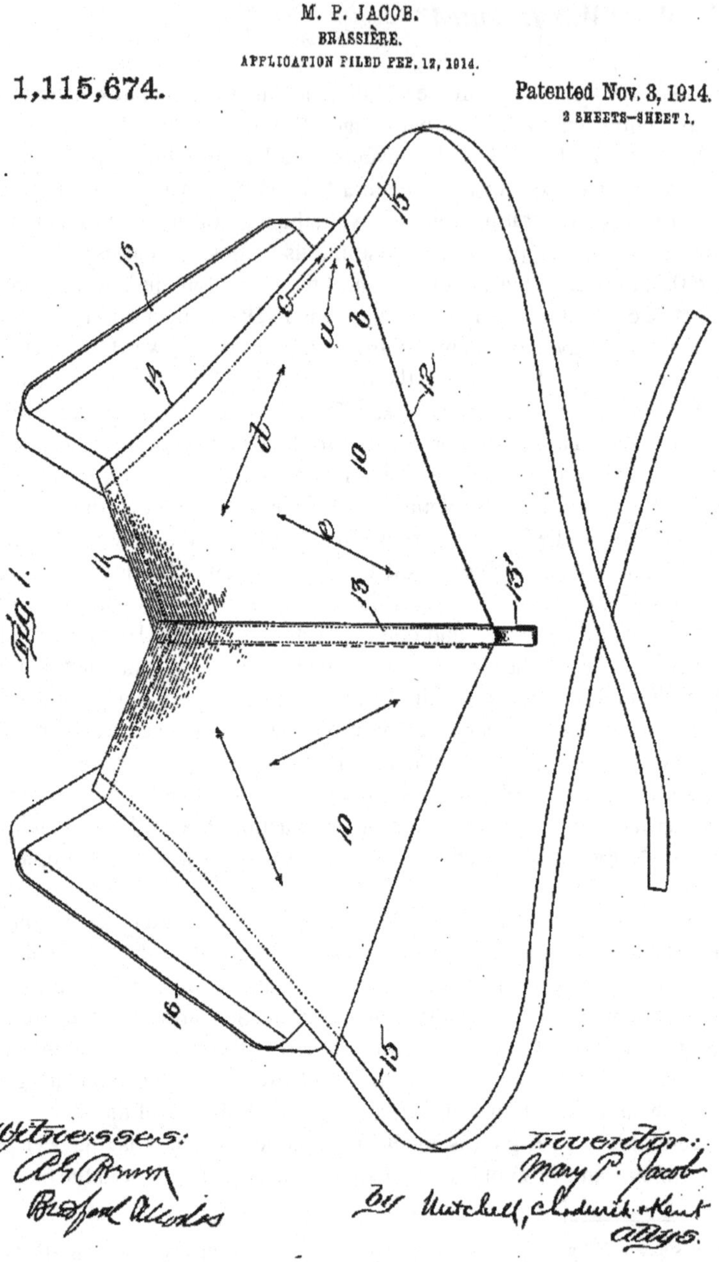

Fig. 10.1 Jacobs backless Brassiere Patent, 1914

started in the early 1920s—the Fashion Form Brassiere Company), but it broke new ground in terms of its design and perhaps was more important in how it helped to change attitudes. In this sense, it is possibly rather like the first smartphone (see *Homo Problematis Solvendis*). It opened up a new line of thinking but was itself quickly superseded by better designs.

10.1.2 Why Was the Brassiere Invented?

There are several candidates for the basic problem here. As has often been the case, there is a more proximate, or immediate, problem as well as a broader problem. The former might be, for example, *how to support breasts*. The latter, on the other hand, might be how to provide comfort or even *how to dress fashionably*. There is an important point here for creativity. Some solutions, as we have seen—think of the toothbrush, for example—very clearly solve a single problem. Others, however, can be seen to solve many problems and not just minor variations on a theme. This latter quality is an important aspect of genesis, as we should be able to observe when we measure this creativity of the brassiere.

10.1.3 How Creative Was the Brassiere?

Relevance and Effectiveness: Mary Jacob's innovation—two simple handkerchiefs and some ribbon—scores 3 out of 4 for this criterion. Our assessment here examines *fitness for purpose*, and within this we can include not only what it does (however we define that), but also how well it achieves its function. If Mary's problem was to support her bust in a way that did not interfere with the appearance of her gown, then she succeeded and did so well enough for her design to attract wider attention. Not only that, but it was simple and inexpensive. However, we must reserve some of our scoring to account for the fact that it was not particularly popular or commercially successful. I feel that this is not unfairly drawing on hindsight but is more a reflection of the fact that it was, in the end, a rather makeshift solution, with plenty of room for further improvement. In other words, it was functional and adequate for her purposes, but no more.

Novelty: I feel a little mean-spirited with the score for relevance and effectiveness, and as you will shortly see, for elegance. I am also conscious that as a man, I have no experience to draw on as a user! However, let me redeem myself with a discussion of the novelty of Mary's brassiere. Here, I have scored it a 4 out of 4. This reflects the real value and importance of this invention. The execution may have lacked something, but the change in thinking that it represents is outstanding. Not only did the backless brassiere help to define the underpinning problem and highlight the shortcomings of existing solutions (i.e. the corset), but it also showed exactly how corsets could be improved, and immediately showed the impact of that improvement. The backless brassiere not only represented an extension of the corset paradigm in a new direction but was a substantially new concept. We can get a sense that Mary (or Caresse when she made these comments) possibly felt the same about her own invention from the fact that she abandoned the business she set up. She appears to have had little interest in developing it further and was satisfied enough with the initial breakthrough.

Elegance: I have already signalled that elegance would not be an outright strength of this invention. It was by no means poor in this respect, but its highly functional nature suggests that Mary's principle concern was the basic problem and less the execution. We also see, as has been obvious in many other cases, that effectiveness and elegance tend to work together. In Mary's case, the initial solution is good enough, but had room for improvement. What is surprising is that she appears to have had little interest in fine-tuning her solution. Had she done so, it is possible that her design would have been a commercial success. I have scored it a 3 out of 4 against elegance. It is certainly simple and reasonably well executed. It just seems to lack a little of the gloss that would be applied by others in the undergarment business. However, the recurring theme that I detect in Mary/Caresse's story is that she was satisfied with the breakthrough and happy to leave the development and refinement to others. That is quite interesting in terms of the psychology of creativity, suggesting that Mary was possibly very suited to the very front-end of the creative process.

Genesis: I have often pointed out that genesis is where we see the more *radical* nature of some inventions, whereas novelty tends to reflect more incremental qualities. The backless brassiere certainly scores well in this criterion—3 out of 4—without excelling. This might seem strange, given what I have already said about this innovation, but I think the key here is the nature of the problem that is solved. If the backless brassiere was simply a solution to the problem of supporting women's breasts, then it might score better for genesis. It was after all a radical departure from the corset. However, because it seems clear that Mary's objective was also related to the fashion aspects of undergarments and their relationship to the fashions of the day, I find less of a radical nature about this invention. In some ways, in other words, the brassiere is a reasonably straightforward solution, not to the problem of anatomical support, but to the problem of designing an undergarment suited to the backless gowns of the day.

This may seem a little paradoxical. As a solution to the same problem the corset is addressing, the brassiere is perhaps more radical. As a solution to the problems of fashion and comfort, it is perhaps more of an obvious improvement. Nevertheless, it remains as important invention as much for its cultural impact as anything else.

Total: The backless brassiere, designed by Mary Jacobs on a whim, scores a creditable 13 out of 16, placing it just inside the *very high* range of creativity.

10.2 The Radio-Controlled Torpedo (1942)

It is not only fine feathers that make fine birds—Aesop, Greek Fabulist and Storyteller (c. 620–564 BCE)

Next, we turn our attention to one of the most paradoxical, and multi-talented, of our *Femina Problematis Solvendis*: the actress and inventor Hedy Lamarr. What makes Lamarr's story so compelling is the diversity of the areas in which she excelled. By day, so to speak, she was a successful Hollywood actress. Indeed, at her peak during World War 2, she was starring alongside actors such as Clark Gable and James Stewart and often was billed as the most beautiful woman in the world. However, it was this tendency to cast her in roles that drew mainly on her beauty and physical allure, rather than dialogue, that bored her and contributed to her seeking intellectual stimulation through problem-solving and invention. Although lacking formal training, her intellect, as well as her knowledge of weapons and munitions, gained through her first marriage to Austrian industrialist Friedrich Mandl, and her concern over the progress of the war, created a set of circumstances ripe for the kind of serendipitous creativity that physicist Ernst Mach (also an Austrian, as it happens) described in 1896 when he wrote about the *part played by accident in invention and discovery.*[4] These circumstances are also what Louis Pasteur was alluding to in his famous statement that "chance favours only the prepared mind". This is not to say that Lamarr's most famous invention was simply blind luck. Rather, *chance* here emphasises that it resulted from a fortunate combination of factors: she was the right person, with the right knowledge, in the right place, at the right time. Let us now look more closely at her invention of a unique form of information-handling system: a secret communication system.

10.2.1 What Was Invented?

Lamarr's innovation, developed in partnership with composer and pianist George Antheil, was granted a patent in the USA in 1942 and was entitled a *secret communication system*. It is now regarded as an example of the application of *spread spectrum technology*. The invention arose from the fact that Lamarr (Fig. 10.2) became aware of the possibility that radio-controlled torpedoes could be jammed easily; i.e. the radio signals controlling the torpedo's direction could be interfered with by the enemy. She hypothesised that this jamming could be avoided if the controlling radio signals, and the receiver on the torpedo, could jump, or *hop*, across different frequencies, thus

---

[4]Mach, E. (1896). On the part played by accident in invention and discovery. *The Monist, 6(2)*, 161–175.

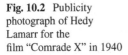

Fig. 10.2 Publicity photograph of Hedy Lamarr for the film "Comrade X" in 1940

eluding the attempted jamming. Lamarr did not invent the concept of frequency-hopping, but what she did invent was the practical application of this concept to a real communication system.[5]

Lamarr and Antheil together succeeded in developing what we might today refer to as a *concept demonstrator*. Not exactly a prototype, which is normally an early version of the intended product, but practical proof that the technology is possible. Using the frequency-hopping concept devised by Lamarr, George Antheil succeeded in synchronising the control of miniaturised *pianola* (or player piano) mechanisms with radio signals. If this seems like a rather odd way to demonstrate the concept for a torpedo control mechanism, by an unlikely person, keep in mind that Antheil had, in 1924, composed a score for a film called *Ballet Mécanique*. Among other things, this involved mechanically synchronising 16—you guessed it—pianolas! *Knowledge* is always an important precursor to creativity.

The secret communication system devised by Lamarr, and proved by Antheil, used the paper scrolls that control a pianola to hop across 88 radio frequencies (88 also being the number of keys on a piano keyboard, and therefore the number of different *notes* that can be encoded onto the scroll). The jam-proof radio control of torpedoes

---

[5]Image Credit: Public Domain (published in the USA between 1924 and 1977 without a copyright notice).

using this method would then be achieved by a controlling aircraft sending radio signals to a torpedo, with both aircraft and torpedo equipped with the synchronising pianola scroll mechanism. Think of it like this: both aircraft and torpedo play the same *tune* at the same time. Every time a note in the song changes, the frequency of the radio-controlled signal changes.[6] Any German listening, and wishing to jam the control signal, must try and guess what the radio frequency is. If they guess incorrectly, they cannot *hear* the control signal (the musical note), and if they don't know the song, they cannot possibly hope to guess what note comes next (i.e. what the next radio frequency will be).

Lamarr's invention, and Antheil's handiwork, was patented in 1942, but not actually implemented by the US Navy during the war. It would not be until the late 1950s that engineers would develop a more compact and practical *electronic* version of this control mechanism.[7]

10.2.2 Why Was the Radio-Controlled Torpedo Invented?

Hedy Lamarr's goal was very clear. As a refugee from the war in Europe, she was keenly aware of the titanic conflict that was shaping the world. She had a creative mind and an interest in weapons and munitions, developed during her marriage to Friedrich Mandl. When she learned that the new technology of radio-controlled torpedoes could easily be jammed, she applied her knowledge and skills, deliberately, in order to solve the problem of *how to prevent jamming?*

10.2.3 How Creative Was the Radio-Controlled Torpedo?

Relevance and Effectiveness: This invention is a little tricky, because we have to be clear about exactly what was *invented*. The concept of frequency-hopping already existed, and radio-controlled torpedoes already existed. Lamarr and Antheil didn't actually invent a working jam-proof, radio-controlled torpedo. What they created was a working demonstration of a jam-proof, mechanically synchronised, frequency-hopping control mechanism that could be applied to torpedoes! This takes nothing away from the invention but shows that this is a rather complex example of innovation. In terms of relevance and effectiveness, they demonstrated its feasibility in a *laboratory* setting. However, we know that it was not adopted by the US Navy

---

[6]These control signals, incidentally, are used to steer the torpedo onto its target. Many torpedoes at this time were unguided and might easily miss their target if mis-aimed.

[7]I recently (November 2019) came across a post on social media, by what looked like a genuine and reputable "magazine", claiming that she invented a "sonar sub-locator" which helped win the Battle of the Atlantic. This is nonsense. For me, it is not necessary to invent a fantasy version of her story—the truth is just as impressive!

because of practical concerns that it would be difficult to implement in actual torpedoes. This has to limit our evaluation somewhat. The reality is that the invention worked, but was not entirely fit for purpose, resulting in a score of 3.5 out of 4.

Novelty: Most of the inventions we are considering were implemented, and that makes it easier to judge their creativity. Somebody actually made the artefact in question, so that it was more than just an idea. I've previously commented on the problem of describing Leonardo da Vinci as a great inventor, because the fact is, most of his ideas were never actually built. Am I a great inventor just because I *think* of the idea of a flying car? Or would you want to see one in action before you congratulated me on my genius?

Lamarr is certainly *not* such an example. My point is that her secret communication system represents an early, but still practical, stage of invention. She had to clearly articulate a problem and generate a possible solution. That solution then had to be evaluated and tested—with Antheil's help—in order to prove the concept. All of this demonstrates outstanding novelty. The innovation explicitly addresses a weakness of existing solutions (they can be jammed), and the concept demonstration with pianola scrolls shows the likely effect of the proposed changes (the control signal is protected from jamming). Lamarr's innovation extended the concept of radio control in a new direction and would also suggest different ways of applying the same frequency-hopping concept (e.g. for securing voice communications over radio). All of these qualities confirm that Lamarr's invention was extremely strong in the various elements of novelty, 4 out of 4 in this category.

Elegance: Here, we need to acknowledge not just the elegance of the narrower invention (the pianola-based control mechanism), but also the intended practical application. Lamarr and Antheil were not intending to stop with the successful demonstration of the control concept. They wanted to develop a practical, jam-proof torpedo. The problem is that their control mechanism was not entirely suited to the intended application. As well as it may have worked, it was not convincing enough to persuade the US Navy to put it into actual use. There were enough concerns over the ability to put this technology into actual torpedoes and controlling aircraft that the US Navy chose to pursue other avenues. When a more practical technology became available after the war—the transistor—then Lamarr's invention became feasible. Like many of our inventions with a weakness in elegance, the issue was not sloppy execution, but execution limited by the technology of the day. I give it 3 out of 4 for elegance.

Genesis: We conclude Hedy Lamarr's invention with a strong score of 4 out of 4 for genesis. Mirroring its strength in novelty, this innovation opens up new perspectives on how the concept might be applied in other areas. It offers ideas for solving unrelated problems: it is not just a means for preventing jamming of torpedoes, and not even simply a way to make communications secret, but is more generally a way to improve the quality and integrity of communications and control systems. It motivates us to find other technologies that can implement the same concepts (solid-state electronics, rather than mechanical components). Lamarr's invention turned a theoretical concept into a practical technology, opening up a new range of perspectives and potential.

Total: With a total of 14.5 out of 16, Hedy Lamarr's frequency-hopping control system concept is very near the top of our catalogue of innovation. The only real area for improvement in this case was the execution. If Lamarr and Antheil had been able to demonstrate not just that the system worked, but that it worked in a real torpedo, it is likely that this invention would have had an immediate impact on the course of naval warfare in WW2. That it was not implemented for this purpose until years later takes nothing away from the very high level of creativity embodied in this invention.

10.3 The Compiler (1952)

Almost everything we know about good software architecture has to do with making software easy to change—Mary Poppendieck, American Author and Programmer

The last of our inventions in the Modern Age is an important precursor to the Fourth Industrial Revolution that is rapidly taking hold of our lives in the twenty-first century. Chronologically, this innovation falls just into the *next* of our epochs (the Space Age), but in technological terms, it represents a transition from the first days of the digital computer—ENIAC[8] completed in 1945—to the kinds of desktop machines that are now so familiar to us. For this reason, I am presenting the compiler, developed by Grace Hopper, as the culminating innovation of the Modern Age.

Although we are now experiencing the huge potential of digitalisation, artificial intelligence, machine learning, and the Internet of things (IoT), these advances have only been possible thanks to the pioneering developments of computer scientist and US Navy Officer Grace Hopper. To understand the significance of her invention, we can use her own words. "It's much easier for most people to write an English statement than it is to use symbols", explained Hopper. "So I decided data processors ought to be able to write their programs in English, and the computers would translate them into machine code".[9] Grace Hopper understood that for general-purpose computers to achieve their potential, it was vital that humans be able to communicate easily with them. Think back to Ada Lovelace's invention (described earlier in this book): she was restricted to converting her instructions into patterns of holes on a punched card. Not very user-friendly! Grace Hopper's innovation was to create a mechanism that could translate the language of programmers (English, in this case), into the language of digital computers (machine code—typically binary numbers).

---

[8]Electronic Numerical Integrator and Computer (ENIAC) was the first electronic, general-purpose computer, and was developed at the University of Pennsylvania.

[9]Source: "The Woman Who Spoke to Computers". The Attic. https://www.theattic.space/home-page-blogs/2019/5/31/the-woman-who-taught-computers-to-listen.

Fig. 10.3 Grace Hopper at UNIVAC keyboard, c1960

10.3.1 What Was Invented?

Grace Hopper was a PhD-qualified mathematician when she joined the US Navy Reserve in 1943. By 1944, she was working as part of the team responsible for the development of the *Automatic Sequence Controlled Calculator* (ASCC), also known as the Harvard Mark I computer. This device was a general-purpose electromechanical computer used late in WW2.[10] By 1949, she had joined the Eckert–Mauchly Computer Corporation (EMCC) (founded by the people who had developed ENIAC) and worked on the development of UNIVAC I. This machine was the first general-purpose, digital business computer (see Fig. 10.3). It was during her time at EMCC that Hopper began to develop a system that could convert English language

---

[10]Readers who have seen the 2014 film *The Imitation Game* will be familiar with this type of computer. The machine depicted in the film was developed by Alan Turing and used at Bletchley Park in the UK to break German codes. '

instructions into machine code (the *bits*—ones, and zeroes—that digital computers process).[11]

As you might expect, tackling a problem that had not previously been addressed met with some scepticism. Hopper was told that her idea was impossible. Computers don't understand English! Fortunately, Grace Hopper was undeterred and continued to work on the problem, firm in her belief that a compiler could be developed that would translate English language instructions into machine code. Hopper first developed her *A Compiler* in 1951 and 1952. This program was, technically, more *linker*[12] than compiler in the modern sense: for our purposes, a key step on the path to a full compiler. By 1952, she had a system that converted mathematical notation into machine code (so, a compiler of *limited* ability). Her goal, however, remained a true compiler which would allow a programmer to enter instructions in English. In the latter half of the 1950s, Grace Hopper worked for the Remington Rand Corporation (which had taken over EMCC) and it was here that she led the team that developed FLOW-MATIC, the first English-like data processing language.

There is an important distinction to be made here. An *ideal* compiler would be able to translate literally any instructions given to it in English (or, of course, German, French, Chinese, Arabic). In practice, this is (or was) perhaps rather ambitious. This may well be achieved in future (and we are, no doubt, some way along this path), but in practice, the compromise that Hopper developed was to create an *English-like* programming language. What this means is that the compiler can only translate a finite set of plain language instructions. It *knows*, in other words, certain terms like ADD or SUBTRACT, and can convert these into machine code, but if you attempted to give it an instruction like WOBBLE,[13] it would have trouble.

Grace Hopper's work, and published papers, in the early 1950s set out the concept of a compiler and led to the development of the *A Compiler.* Her work in the latter half of the 1950s led to the development of the language FLOW-MATIC. The culmination of this work came in 1959 with a two-day *Conference on Data Systems Languages* (CODASYL). It was here that Common Business-Oriented Language (COBOL), combining elements of FLOW-MATIC with an IBM language known as COMTRAN, was first defined. COBOL remains widely used around the world, especially in large-scale government systems and in fields such as banking, insurance, and health care.

10.3.2   Why Was the Compiler Invented?

When I analysed Ada Lovelace's computer program, I suggested that the problem she addressed was *how to program computers.* That could be an appropriate description of the problem here as well, except that it is probably a little too general. The issue that

---

[11]Image Credit: Flickr: Grace Hopper and UNIVAC. Creative Commons 2.0: https://creativecommons.org/licenses/by/2.0/legalcode.

[12]A computer utility that combines object files generated by a compiler into a single executable file.

[13]If WOBBLE is actually an instruction in FLOW-MATIC or COBOL, then I retract my statement!

Grace Hopper was tackling was not just any way of programming a computer (which could be punched cards or machine code), but a *user-friendly* way of programming computers, a way that is *intuitive* for humans. In fact, I am going to stick with the same problem statement, but with the knowledge that we now have a much clearer idea of not just the *what* of the problem, but also the *how well* it should be achieved. Grace Hopper's problem was no longer about getting machine-readable instructions into the computer but about getting the computer to understand *human-readable instructions*!

10.3.3   How Creative Was the Compiler?

Relevance and Effectiveness: Taking FLOW-MATIC as the actual solution (this being the first solution to the problem of programming in human-readable (i.e. English) instructions), I give Grace Hopper's invention a score of 3.5 in this first category. Although it was entirely feasible with the technology of the time, I feel that we have to be rather strict and acknowledge that it did not achieve the lofty goal that Hopper herself set only a few years earlier. That is no criticism of Grace Hopper: she was absolutely right to shoot for the top. Nevertheless, there must have been instances with FLOW-MATIC where users were frustrated that it couldn't do what they wanted it to do. It was, in other words, limited to a set of predefined English instructions. Even 3.5 is a little generous.

Novelty: Conversely, Hopper's compiler/language FLOW-MATIC was undeniably novel. I'll depart from my more complex definitions of creativity for a moment (see Appendix A) and go back to a more intuitive definition for novelty. When a solution is new, or original, or surprising, then it is novel. When it elicits a reaction like "wow, I've never seen that before!", then it is novel. Grace Hopper was told that her goal of a programming language using English language instructions was not possible. She created a solution that proved the doubters wrong, and when FLOW-MATIC is described as the first English-like programming language—when you've never seen something like it before—then it is, by definition, *novel*! It deserves the maximum score of 4 in this category.

Elegance: When you look at some code written in FLOW-MATIC, even if you know very little about programming, you get a sense that it is well executed. You can see the careful structure implicit in the example of FLOW-MATIC code shown below[14]:

(1) TRANSFER A TO D.
(2) WRITE-ITEM D.
(3) JUMP TO OPERATION 8.
(4) TRANSFER A TO C.

---

[14]This example is taken from Sammet, Jean (1969). *Programming Languages: History and Fundamentals*. Prentice-Hall, pp. 316–324. ISBN 0-13-729988-5.

Equally, it is clear that the instructions mean something in English and are designed to get the computer to perform related operations. In order to function as a programming language, FLOW-MATIC had to be convincing, complete, and fully worked out. A half-complete solution, in other words, one in which some instructions didn't execute or in which random words might be executed as instructions, would be unacceptable. Therefore, in isolation, FLOW-MATIC was a highly elegant solution and scores 4 in this category. If it had only a limited vocabulary, that is not a matter of elegance, but of effectiveness.

Genesis: Finally, we must ask if Grace Hopper's invention had an impact beyond the immediate problem? What was its impact on the external world? How did her concept of a compiler, and a programming language using English language instructions, change how we understand the problem of how to program computers?

There can be little doubt that Hopper's innovation was *foundational*. It has been the basis of decades of further work, refining and improving on the basic solution concept. It was also *seminal* in drawing attention to previously unnoticed, or unknown, problems. For example, what English words must the compiler understand? What English instructions are necessary for a useful language? Grace Hopper's invention broke a fundamental paradigm of computer programming, embodied in the statements of colleagues who said it couldn't be done. The naysayers said that computers cannot understand English, but Grace Hopper proved that they could, and her invention is worthy of a score of 4 for genesis.

Total: With a total of 15.5 out of 16, Grace Hopper's invention of both the *idea* of programming in natural language and the *means* to do so is just shy of a perfect score. The only minor criticism relates to effectiveness, and that is only the acknowledgement that there is a theoretical ideal. This aspect of computer languages would improve over time and may yet improve further, but Hopper retains the credit for transforming computer programming and helping to pave the way for the Digital Age that we now live in.

Chapter 11
The Space Age (1950–1981): The Rise of the Modern Creativity Era

The penultimate period that we will tackle is the *Space Age*. For me, born in 1967, this still looms large as an age of exciting, grand achievements: landing humans on the Moon; the Space Shuttle; and space craft sent to the outer reaches of the solar system. I continue to find it thrilling to read about the Apollo program and the ingenuity with which NASA tackled multiple technological challenges using rudimentary digital computers and other equipment that seems almost laughably simplistic to a twenty-first-century eye. The computer on board the lunar landing module of Apollo 11 had 2 k of memory and ran at a speed of 1 MHz (compared to a typical modern smartphone with a clock speed that is more than 1000 times faster and usually at least two or three million times as much memory). I have a vague memory of watching the 1969 Apollo 11 Moon landing on a grainy black and white TV, and throughout my childhood the epithet "Space Age" was applied to anything that people wanted to show was supermodern. So, remember kids: those *boomers* you make fun of were the people who developed the computers and smartphones you depend on!

Although we are considering inventions of the Space Age, we won't see anything space-like in this chapter. We begin with a very down-to-earth invention. Something we have all used and something that most parents are grateful for is: the *waterproof diaper* (or nappy). What is interesting about this invention is that you probably would not think of diapers/nappies as something that is both a hotbed of innovation and also very big business. In fact, there is a connection to the Space Age, because the Apollo astronauts wore adult nappies when they went on their lunar excursions (and left the used ones on the Moon).

The second innovation in this era is rather more high-tech than the waterproof diaper. *Kevlar* is a synthetic fibre, only possible thanks to modern advances in chemistry, oil production, and automotive transportation. In other words, it is not something that could ever have simply been discovered and adapted. Kevlar shows us another facet of modern innovation, and that is what happens when efforts to invent are scaled up and systematised. In other words, Kevlar shows us something of the deliberate, scientific, and industrial approach to creativity and innovation.

© Springer Nature Singapore Pte Ltd. 2020
D. H. Cropley, *Femina Problematis Solvendis—Problem solving Woman*,
https://doi.org/10.1007/978-981-15-3967-1_11

We close this epoch with another invention that is rather similar in origin to Kevlar. Like Kevlar, *Scotchgard* resulted from an industrialisation of creativity, innovation, and problem-solving. Armed with highly qualified scientists, novel materials, and a deliberate process of seeking out profitable problems to be solved, the company that was responsible for the development of Scotchgard was able to create an environment that maximised the likelihood of finding novel and effective solutions. So much so that many other organisations today try to copy the approach and company culture that gave rise to inventions such as Scotchgard.

11.1   Waterproof Diaper (c1949)

The Conservative establishment has always treated women as nannies, grannies and fannies.—Theresa Gorman, British Politician and Author (1931–2015)

The first invention that we consider in the Space Age might seem almost as far as it is possible to be from rockets, astronauts, and the Moon! However, even the Apollo astronauts in the 1960s required a solution for dealing with their bodily functions. Part of the so-called *suiting-up sequence* for Apollo astronauts' excursions on the lunar surface included not only a *urinary collection subassembly*—a complex system of tubes worn under the astronauts' clothing, for collecting and disposing of liquid waste—but also the impressively named *faecal containment system*. This latter "system" was little more than a giant man-nappy.

We don't have to be space travellers to appreciate the importance and utility of any invention that helps to deal, hygienically and conveniently, with the problems associated with our bodily functions. In *Homo Problematis Solvendis*, I described the importance of the invention of modern sewer systems, and the modern flush toilet has also been an important contributor to the development of healthy societies. It is one thing, however, for independent, healthy children and adults to make use of toilets, but what solutions do we have for babies or even adults, who are unable to control their bodily functions? Humankind has probably used nappies/diapers in some basic form for centuries. While they began to take on their modern form only in the nineteenth century, these examples only solved part of the problem. An important innovation, in the late 1940s, was to make them *waterproof*, and American inventor Marion Donovan was responsible for this material-handling invention.

11.1.1   What Was Invented?

Marion Donavan had a problem that mothers, and fathers, have faced for millennia. Our human offspring begin their lives in a very helpless state. Most parents will joke that newborn babies spend most their time sleeping, feeding, or excreting, and this a pretty accurate description! Human infants have no control over their bladder or bowel for the first 12 months of their lives. From the age of about 12–18 months, they begin to have some control, but it is not until somewhere in the range 24–30 months that children learn to control these parts of their bodies. The average age at which our children are successfully toilet-trained is 27 months.

This means, of course, that infants require constant help with these functions. That's a polite way of saying that parents spend a great deal of time cleaning up the results of our babies' *accidents*. In ancient times, our distant ancestors may have had a different approach to infant hygiene. Nowadays, however, the way we live our lives, in comfortable homes, with nice clothing, comfortable beds, and a much greater concern for, and awareness of, hygiene, means that it is inconceivable that infants might simply do their business anywhere that takes their fancy.

Marion Donovan, like parents before and after her, used cloth nappies (diapers, if you are from North America) with her infant daughter. The problem with cloth nappies (my wife and I briefly flirted with them, so I can vouch for this) is that they really only *catch* the mess. They absorb some urine and other liquid waste, but also have a tendency to leak once saturated. Leaking means that clothing and bedding get soiled, and before you know it, one innocent bowel movement means a full washing machine. Marion therefore attacked this problem head-on, turning to her inventive abilities to find a simple remedy. The solution was simple: cover the cloth nappy with a waterproof material (see Fig. 11.1), and the waste might be both caught *and* contained.[1]

In fact, Donovan first used a plastic shower curtain to achieve this. This itself is an interesting example of an element of creativity, namely the ability to form *remote associates*. What I mean by this is the ability to see that a shower curtain is not just a way to stop your shower water making a mess of your bathroom floor but also more generally a waterproof cloth. Like any cloth, it can be cut, and sewed, and made into other wearable items. Her breakthrough was not only the idea that nappies would be better if they didn't leak, but that a light, flexible, waterproof cloth has many different applications.

The waterproof diaper would undergo many further improvements. Most of these can be seen as incremental in nature: how to improve their capacity to not leak, how to make them easier to put on and take off, and so forth. Even over the period that my own children used diapers—roughly eight years spanning the late 1990s and early 2000s, unless they aren't telling me something—there have been significant changes with highly absorbent, biodegradable, stretchy, gender-specific, scented disposables the norm. For now, let us consider Marion Donovan's breakthrough and define the underlying problem, before we consider how creative this invention was.

---

[1] Image Credit: http://www.uspto.gov, Public Domain. Source: https://www.smithsonianmag.com/innovation/meet-marion-donovan-mother-who-invented-precursor-disposable-diaper-180972118/.

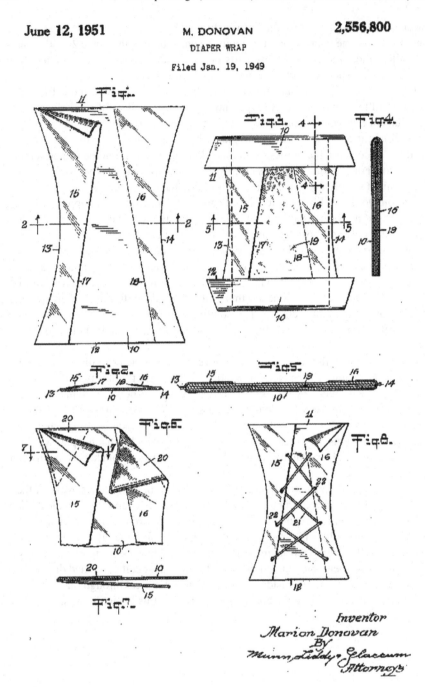

Fig. 11.1 Donovan's Waterproof Diaper Wrap Patent, 1949

11.1.2 Why Was the Waterproof Diaper Invented?

Marion Donovan's frustration with cloth nappies was based on a very clear problem. Although they did a good job of catching her infant daughter's efforts, they had a limited ability to absorb liquid. Once full, they leaked. In fact, even before they were full there was a risk that they could leak. The problem was simple enough: *how to stop leaks*. In some of our other inventions, I have already allowed us a little leeway in these *how to verb noun* definitions. It is implicit that the focus is nappies/diapers, so we don't need to state that explicitly. That's handy, because it makes our job of finding a four-word problem statement a little easier. Similarly, I've also introduced the idea that we can supplement our basic definition with additional information, usually regarding *how well*? This constraint is important, because it also gives us a mechanism by which we can judge the effectiveness of solutions, once we get to the analytical stage of problem-solving. In this case, how well (in engineering language, a *non-functional performance requirement*) is probably *completely*.

11.1.3 How Creative Is the Waterproof Diaper?

Relevance and Effectiveness: Marion Donovan's waterproof diaper was entirely feasible, using the materials, knowledge, and technologies of the day. This quality—which I define as *correctness*, in Appendix A—captures the fact that an invention is really most useful when it is not just a great idea but a great idea that can be implemented. This notion is captured very neatly in the recent Australian documentary *2040*, which explores solutions to climate change and global warming. A key concept in the film is that the solutions considered must be *correct* in our sense. They must exist now and must be capable of implementation now.[2]

Back to our waterproof diapers, which are strong in correctness, but a little less impressive in performance and appropriateness. These latter qualities test the solution's ability to do what it is supposed to do, and how well it achieves this. Donovan's nappies addressed the basic problem but did so imperfectly. Having used cloth nappies and waterproof covers, I can confirm that the weakness of Donovan's design is that there are still gaps in the waterproof barrier. The waterproof cloth design can still leak around the legs and waist. Not only that, but the absorptive capacity of the nappy is unchanged. It still gets saturated fairly quickly and therefore will still leak, just not as badly, or quickly, as without the waterproof cover. Despite these drawbacks, I will still give it 3.5 out of 4 for effectiveness. Like all of our inventions,

---

[2]The point of the documentary, of course, is that there are many such solutions to the problems of global warming and climate change. In fact, they are not only *correct*, but also strong in *performance* and *appropriateness*. Together, that seems to make them highly *relevant and effective*, aside from any other characteristics of creativity that they might possess. If they aren't being implemented, that suggests that the blockage lies elsewhere. Presumably, politics!

they wouldn't be in this book if they didn't work fairly well. Perhaps the blame for this slightly imperfect score lies elsewhere?

Novelty: In general, in our assessments of novelty I have stuck to a single notion of originality. However, as you will see in the explanation in Appendix A, there are actually two distinct aspects of novelty. The first is *problematisation*: the extent to which the solution helps to define the problem. The second is *propulsion*: the extent to which the solution sheds new light on the problem. It is convenient, and somewhat simpler, to group these together under the single banner of novelty. However, in this case, I want to deal with them separately. The reason for this is that quite often an invention scores rather differently in these two subcategories, and a single score for novelty masks these differences. The waterproof diaper scores well for problematisation. It crystallises the nature of the problem, what needs to be done, and how it could be done better. Conversely, in the case of propulsion, the waterproof diaper is decidedly, though not fatally, weaker. In this latter case, its main limitations are that it does not suggest too many ideas for new and different ways of using the solution, and it does not really offer fundamentally new perspectives on the general nappy concept. What this says, in plainer language, is that the waterproof diaper that Marion Donovan invented was not, itself, such a great invention, but it does a fantastic job of illustrating what needs to be done. It is like an icebreaker! Once we see this invention, we immediately see how it could be improved. It took Donovan's invention to open our eyes to new possibilities. That, of course, still makes it quite important and creative. As a result, it gets a score of 3.5 for novelty.

Elegance: Under effectiveness, I suggested that the imperfect score in that category might, in fact, be the fault of something else. I think that fault lies, at least partly, with elegance. The waterproof diaper was undoubtedly convincing and pleasing. I am sure that Marion Donovan's sewing skills were such that we would have looked at this invention and though "Oh, that's a neat idea, and nicely done!" However, as well executed as it looked (and was) it must be regarded as slightly incomplete and slightly imperfectly proportioned. This is manifest in the weak elements of effectiveness. A slightly better design would have addressed the possibility of leaking around the legs and waist, for example. Regrettably, I have to give Marion Donovan's waterproof diaper 3.5 out of 4 for elegance.

Genesis: The waterproof diaper gets 3.5 out of 4 in this category. This reflects quite strong qualities across the indicators of genesis. This invention, for example, suggests a novel basis for further work. What we mean here is not simply the idea of stopping the leaking a bit better but stopping it by finding different ways to prevent liquid from escaping. Donovan's approach was a waterproof barrier, but perhaps another approach is simply to make the nappy so absorbent that no liquid can escape? Another quality in this category is germinality: the solution suggests a new way of looking at the problem. In this case, that might be not merely absorbing the liquid, and not even just blocking it from escaping, but instead, locking it in so that it cannot escape. In addition to these qualities of genesis, the waterproof diaper also set a new benchmark for nappies. From this point on, a good nappy was not merely one that caught the mess, but one that contained the waste. This was the jumping off point for many further innovations in diaper design. When you consider that more than 27 billion nappies are sold in the USA alone each year, you can see why this is an important area for innovation.

Total: Despite being not quite perfect in every category, the waterproof diaper invented by Marion Donovan scores a very healthy 14 out of 16 for creativity. This *very high* score has only minor areas for improvement in each of the categories we consider. Its performance could have been slightly better. Its propulsion, as a component of novelty, slightly stronger. If it had been just a little more complete and well executed, it would have been both more elegant and more effective. Finally, with just a little bit more that broke the paradigm of conventional nappies, we would be looking at another invention with maximum creativity.

11.2 Kevlar (1964)

My faith protects me. My Kevlar helps—Jim Butcher, American Author (1971–)

Some years ago, I needed some dental work done. I cannot remember exactly what, but I do remember being surprised that my dentist opted to use a small piece of Kevlar as part of the treatment. What caught my attention was the fact that my dentist explained that he had tried to save some money by purchasing the Kevlar, but not the special scissors needed to cut it. He quickly discovered that ordinary scissors were not up to the task.

What really surprised me is the versatility of this material. Like many people, I only associated Kevlar with bulletproof vests and military helmets, and until the dentist embedded a small piece in me, I had no idea of the range of applications to which this synthetic fibre could be put. Although Kevlar is a fibre—in effect, like a thread or yarn—and can be woven into fabric, I do not consider it a material-handling system, but an energy-handling system, because of its key properties and the uses to which it is put.

Kevlar is a good example of the way in which modern technologies, such as the twentieth-century developments in synthetic and polymer organic chemistry, opened up all sorts of incremental improvements to existing solutions. Stephanie Kwolek's innovation, originally targeting the development of lightweight car tyres, also demonstrated its value for many other problems.

11.2.1 What Was Invented?

Stephanie Kwolek majored in chemistry at Carnegie Mellon University, in Pittsburgh, graduating in 1946. She took a job with the DuPont company (familiar to us for products such as *Styrofoam*, first discovered in 1941) intending only to use this job to earn enough money for medical school. However, she quickly found herself deeply engaged in her work at DuPont and stayed with the company until her retirement in 1986.

Kevlar is a polymer. Technically *poly-paraphenylene terephthalamide*, in very simple terms, it is somewhat like nylon in concept. However, compared to nylon, Kevlar is extremely strong. In fact, it is around five times stronger than steel by weight. Kevlar begins life as a cloudy solution (i.e. a liquid). This solution is passed through a *spinneret*—basically, a nozzle—and as it squirts out in a fine stream the polymer solidifies, creating a fibre (see Fig. 11.2). This fibre can be further strengthened with some heat treatment and then woven into fabric.[3]

Kevlar has some remarkable properties. Its *tensile strength* (its ability to withstand stretching) is measured at 3620 MPa. For comparison, aluminium is about 300 MPa,

Fig. 11.2 Golden yellow aramid (Kevlar) fibre

---

[3]The image depicts raw golden yellow aramid fibre (Kevlar). The diameter of the filaments is about 10 μm. These fibres are used to produce the woven Kevlar fabric. Image Credit: Cjp24, Creative Commons 3.0: https://creativecommons.org/licenses/by-sa/3.0/legalcode.

carbon steel is 841 MPa, nylon is about 83 MPa, and modern nylon-based ropes are about 1000 MPa. Kevlar's relative density (compared to water) is 1.44. Steel, on the other hand, has a relative density of around 8, while aluminium is 2.7. Finally, Kevlar also maintains its strength at extremely low temperatures (e.g. in the region of $-200\,°C$), and even long periods at relatively high temperatures (e.g. $+160\,°C$) only reduce the strength of Kevlar by about 10%. It is thanks to these qualities that Kevlar has found a wide range of applications, from body armour to car tires to loudspeaker cones.

The story of Kevlar is an interesting blend of invention and discovery. In fact, it is a classic example, and perhaps even an extension of, Pasteur's famous dictum (*chance favours only the prepared mind*) that we referred to back in the discussion of soap. Kwolek was not just *prepared* but was actively searching for what she found. She didn't really stumble across Kevlar. The fact is she was actively engaged in trying to solve a problem and understood the importance of exploring every possible solution (both in the sense of polymer liquids and problems) that presented itself.

11.2.2 Why Was Kevlar Invented?

The development of Kevlar gives us an excellent example of two essential ingredients for innovation success. One is already familiar to us: we need a clear statement of the problem. The second stems from research which has shown[4] that three elements of the work environment are vital for innovation. Innovators need sufficient time, enough resources, and positive support to succeed. What is interesting about these three factors is that there is a *sweet spot* for them. Too little in the way of time, resources and support kill off innovation. However, too much can also be a bad thing. It is rather like Goldilocks: not too hot, not too cold, but just right.

Stephanie Kwolek worked for an organisation, and in a team, that had a clear problem to solve. In response to a looming fuel shortage in the mid-1960s, her group was searching for a new fibre that could be used to make lightweight, strong car tires.[5] The goal was to improve car tires, but the specific problem was *how to optimise polymer fibres*. Working in a well-resourced, supportive team environment, for a company with a strong tradition of innovation, the scene was set for a new invention.

---

[4]For example, Amabile, T. M. (1983). *The social psychology of creativity*. New York, NY: Springer.

[5]The connection to a fuel shortage lies in the impact that tires have on fuel consumption. Larger, heavier tires increase fuel consumption, as does the tire's *rolling resistance*. These tire properties are strongly tied to the strength, mass, stiffness, and other physical characteristics of the tire. In the USA alone, it is thought that as much as 15% of fuel consumption is related to the rolling resistance of car tires.

11.2.3   How Creative Was Kevlar?

Relevance and Effectiveness: The polymer fibre that Stephanie Kwolek created was highly fit for purpose. A problem like *optimising polymer fibres* (in this case, making them stronger and more versatile) is an open-ended goal. In theory, there is no upper limit to this optimisation. This could be an excuse to mark the effectiveness of Kevlar at something less than 4. However, that fact that Kevlar exceeded not just other polymers, but other materials like steel, means that it would be churlish to argue about its feasibility, performance, or its appropriateness. Kevlar scores 4 out of 4 in this category, without a doubt.

Novelty: In simple terms, Kevlar was new and surprising (at least for chemists who understand polymers). In fact, for anyone with a passing familiarity with polymers in the mid-1960s, it must have been a surprising, unexpected development. If your only knowledge of polymers had been nylon, you would have been amazed at the properties of Kevlar.

Even in our more technical assessment of novelty, Kevlar scores very strongly. It clearly highlights the weaknesses of other solutions to the same problem. We only have to consider the broader goal of making lighter, stronger car tires to see this. More narrowly, we can see that it is an outstanding optimisation of polymer fibres. Kevlar is also very strong in terms of propulsion. In particular, it is strong in *redirection* (extending the known in a new direction) and *redefinition* (highlighting new and different ways to use the solution). That seems to be especially so for Kevlar, which has led to an astounding array of applications, often far removed from the original goal of making fuel-efficient car tires.

Elegance: Some solutions to problems are disarmingly simple. Not necessarily in terms of their internal make-up, or chemical formula, but in their external appearance. What could be simpler than a polymer thread that can be spun into a cloth, just like cotton or wool? When that thread has superpowers, of course it leads to some interesting possibilities. Elegant solutions often also have a strong degree of *obviousness* after the fact. There must have been many chemists, who followed in Stephanie Kwolek's wake, who experienced a moment of *why didn't I think of that?* Kevlar is really very simple as a concept. The process of creating it is straightforward and understandable, and the execution is quite simple and routine. As we have seen in many of the inventions in this book, this elegance then typically supports strong effectiveness, and that is the case here. Kevlar gets 4 out of 4 for this category.

Genesis: Our final category, as I have often indicated, is the toughest. Inventions that score well in this dimension of creativity do so because they create new paradigms, make quantum leaps ahead of existing solutions, and fundamentally change our understanding of the problem. This latter quality is sometimes visible when we come across an invention that makes us realise that *the secret to solving this problem is actually xxxx.* That is not a misprint. In creativity workshops, I often give people the task of making a tower out of some sheets of paper. What they often realise, usually late in the exercise, is that the secret to making a paper tower is the problem of *how to make paper stiff*. This is a quality captured by genesis: some solutions cause you to realise that the problem you thought you were solving is really quite different. In the case of Kevlar, we see the end result of this in the enormous range of applications for this high-strength fibre. It solves many problems, not just the original one. Who would have thought, prior to 1965, that a polymer fibre, woven into cloth, could stop bullets just like steel plate? Kevlar gets 4 out of 4 for genesis.

Total: This brings us to a final score of the maximum possible: 16 out of 16. Kevlar, and the circumstances of its invention serve as a great case study of creativity and problem-solving. The invention itself is extremely creative, as assessed by our criteria. It is also an example of what happens in creativity and innovation when all of the key elements align. When a highly knowledgeable, motivated, open, persistent person, working in a supportive environment, and equipped with the ability to generate and analyse a broad range of ideas, is able to apply this competency to a well-defined problem, innovation happens!

11.3   Scotchgard (1973)

Repentant tears wash out the stain of guilt—Saint Augustine, Roman African Theologian and Philosopher (354CE–430CE)

Some years ago, my wife and I bought three beautiful new sofas. In hindsight, this was a bold move, because we had two young children at the time, with a third not far off. These sofas were a lovely grey suede, and they were the first that we bought brand new. Naturally, they took pride of place in our home. It was a bold move, because, as any parent knows, children are messy. Not to be denied our beautiful sofas, we at least had the sense to pay for the special fabric treatment, guaranteed to protect these sofas from all sorts of stains. Knowing me, I probably thought this was a waste of money, but we paid for the treatment and thought little more about it.

Of course, not long after, the inevitable happened and my eldest son—now a writer, as it turns out—scribbled on one of the new sofas with a ballpoint pen! My initial reaction was that this had ruined the sofa and would be impossible to remove. However, I did remember the special fabric treatment, and I had not yet lost the instructions. So, with great hope, I did what the instructions said and gently applied some dish-washing detergent. I rubbed this in carefully, making small circles, before wiping it off with a damp cloth. To my amazement, the ink completely disappeared! This may not have been *Scotchgard* by name, but it was a descendant, and no doubt traced its origins back to Patsy Sherman's pioneering fabric protector, for which she shared the patent (with Samuel Smith) in 1973.

11.3.1   What Was Invented?

Patsy Sherman worked for the company 3M. This organisation is probably familiar to you for a range of products including a variety of reusable *hanging products* (e.g. picture hooks that stick to the wall and are easily removed), sandpaper (the company's first product), and the ubiquitous *Post-it* notes that no self-respecting creativity consultant can be seen without. Renowned for a company culture that has, in their own words, a *tolerance for tinkerers,*[6] 3M has grown from its origin in 1902 to a global organisation with more than 90,000 employees and 60,000 products.

The work that led to the development of Scotchgard began with 3M's move into the manufacture of *fluorochemicals*, in the 1940s. In particular, the company began manufacturing perfluorooctanoic acid, or *PFOA*, and in 1951 began supplying PFOA to the DuPont company for the manufacture of *Teflon* (think non-stick frying pans). Patsy Sherman, a chemist, joined 3M in 1952, working on fluorochemicals.[7] Her first task, with co-worker Sam Smith, was to develop a new kind of rubber that had properties especially suited to use in jet aircraft fuel lines.

In the course of this work, in 1953, Sherman had developed a synthetic latex. One day, a laboratory assistant accidently dropped the bottle containing the synthetic compound, splashing some of the substance on their canvas shoes. Like any good scientist and problem solver, Sherman and Smith were observant and noticed three things about the shoes. First, the synthetic compound did not affect the appearance of the shoes. Second, the compound could not be washed off the shoes. In other words, it appeared to be impervious to solvents. Third, it repelled substances such as water and oil.

Sherman and Smith immediately realised the potential of this new compound. If you've ever been camping in an old canvas tent, you probably know what happens if it rains. One solution is to carry a flysheet—a second, outer layer—for the tent, but this means extra weight. What if you could waterproof the tent with a lightweight, chemical coating? Sherman and Smith, and the company itself, realised that their PFOA-based compound had all sorts of applications beyond mere waterproofing. If liquids and oils, for example coffee, gravy, sauce, are repelled by this compound, then it follows that materials such as upholstery and carpet, impregnated with the chemical, will not be able to absorb substances that would otherwise soak in and stain them (see Fig. 11.3). Thus, Scotchgard—the fabric protector—was born.[8]

Over the course of the next two decades, 3M developed a range of products based on the properties of this *material-handling* compound. However, the story did not

---

[6]See *A Century of Innovation: The 3M Story*: http://multimedia.3m.com/mws/media/1712400/3m-century-of-innovation-book.pdf.

[7]Fluorochemicals, simply chemicals derived from the element fluorine, are big business. Coming in various forms—*fluorocarbons* (some of which, like CFC, were phased out after their ozone-depleting properties were discovered), *inorganic fluorides* (used for water fluoridation), and *fluorine gas* (used in uranium enrichment)—some 3 million metric tonnes are produced each year.

[8]Image Credit: https://www.marlentextiles.com/feb2017.html, Public Domain.

Fig. 11.3 Water-resistant fabric treated with Scotchgard

quite end there. In 2000, in response to concerns about the toxicity of some fluoro-chemicals, in particular PFOA and *perfluorooctane sulphonate* (PFOS), 3M phased out production of these compounds and in 2003 reformulated Scotchgard using a different fluorochemical with a much shorter half-life in humans (approximately one month as opposed to over five years: this is the time it takes for the chemical's concentration in the blood plasma to reduce by half[9]). More recent concerns about the chemicals used in products such as Scotchgard should not detract from the creativity of this invention. The story of Scotchgard does, however, remind us that creativity, innovation, problem-solving, and invention do not take place in a consequence-free environment. The history of creativity is filled with examples of unintended consequences and deliberate misuse. Ethical considerations should always guide our problem-solving efforts (for a deeper discussion, see the book by Moran, Cropley, and Kaufman, 2014, listed in the bibliography).

11.3.2 Why Was Scotchgard Invented?

The story of Scotchgard is often told in the same way as Kevlar. Both, it is often said, are examples of *discoveries* rather than deliberate invention. I tend to disagree

---

[9]The logic here is that the faster a substance is removed from our blood plasma, the less harm it can do. The scientific consensus seems to be that there is *no conclusive evidence* that these substances cause harm to humans. However, it seems to be a wise precaution to minimise the potential risk.

and see very few instances of genuine, serendipitous, lucky discovery. One way to differentiate between invention and discovery[10] is to consider which came first: the problem or the solution? Discoveries can be thought of as pre-existing solutions that were waiting for a problem, whereas inventions are solutions developed in response to a problem. Even then, I think it is rare to find a discovery that was not, in some way, a deliberate creation, even if the problem wasn't quite what was expected. Alexander Fleming's "discovery" of penicillin is a case in point. The only reason he *discovered* this substance was because he was trying to solve a related problem (he was investigating the antibacterial properties of nasal mucous, in fact). The reason Fleming, and not the cleaner, discovered penicillin, is that Fleming had what Mach described as *the capacity to profit by experience*.

So it is with Scotchgard. Patsy Sherman (and Samuel Smith) *discovered* this compound because they were engaged in a problem-solving process. The chemicals in their synthetic latex didn't accidently fall into a beaker and mix themselves. Unlike fire, fluorochemicals don't occur naturally. You don't wander the woods and accidently come across a pool of PFOA. Sherman and Smith had the capacity to profit from their experience because they were observant and prepared, and because they were *looking*. The only thing that makes this seem a little like discovery is that their solution solved a different problem. I have no hesitation saying that Sherman and Smith *invented* Scotchgard, just as deliberately as Hedy Lamarr invented a radio-controlled torpedo, or Hildegard of Bingen invented Lingua Ignota. It didn't take Sherman and Smith, or 3M, long to understand that they had a solution to the problem of *how to protect fabric*.

11.3.3   How Creative Was Scotchgard?

Relevance and Effectiveness: When you read accounts of the development and properties of Scotchgard, you often see words like *miraculous* and *magical*. Just spray it on your sofa or carpet, and these things are protected by an *invisible shield*. What is striking about Scotchgard is that it really worked, as I discovered with my sofa. The rhetoric really seemed to match the reality. Scotchgard was technically feasible and, indeed, could be manufactured in a cost-effective, consistent manner. It did exactly what it was supposed to do: it protected all sorts of fabrics in all sorts of forms, from liquids and oils, preventing staining and helping to keep these fabrics clean. If all we considered were correctness and performance, then it would score the maximum in this category. However, we also have to consider *appropriateness*. As we have seen many times, this calls into consideration some of the constraints on the solution. If performance is *what* the solution has to do, then appropriateness is *how well* it does

---

[10]If you are interested in a deeper discussion of this, I've written a discussion based on Ernst Mach's (of *Mach number* fame) 1896 paper *On the Part Played by Accident in Discovery and Invention*, which he published in *The Monist*. See: Cropley, D. H. (2019). Do we Make Our Own Luck? Reflections on Ernst Mach's Analysis of Invention and Discovery. In V. P. Glaveanu (Ed.), *The Creativity Reader*. Oxford, UK: Oxford University Press.

it. This must include not just *from all possible liquids*, but also *without causing harm.* The concerns about PFOA and related fluorochemicals that surfaced in 1999 throw a little doubt on this aspect of the original Scotchgard invention. Enough, in my mind, to knock the score down to 3 out of 4 in this category.

Novelty: Scotchgard is strong in novelty. In particular, it helps to define the exact problem of protecting fabrics from dirt and stains. What I mean here is the difference between an old-fashioned dust sheet and Scotchgard. The former defines the protection problem as a matter of stopping liquids and dirt from coming into contact with the fabric. The latter defines it as a matter of preventing liquids and dirt from attaching themselves to the fabric. The weakness of a dust sheet, or even a clear plastic cover, is that it changes a carpet or a piece of furniture. To protect these objects, the dust sheet or plastic cover requires you to give up other properties of the object. The dust sheet, for example, costs you the aesthetic appearance, and the *feel*, of your sofa. Scotchgard changed this by allowing liquids and dirt to come into contact with the object you are protecting. This then becomes both an extension of the protection idea and a radically new approach. The protective layer operates at a microscopic level, and the underpinning basis (repelling liquids, oils, and dirt) can keep fabrics clean, prevent stains from permanently damaging the fabric, and can waterproof fabrics as well. Scotchgard is worthy of 4 out of 4 for novelty.

Elegance: I have frequently repeated an engineering adage that says a good solution usually *looks like* a good solution. This is a rather macroscopic view of elegance. If we also apply the same principle at a more microscopic level, then convincingness, pleasingness, harmoniousness, and the like should also feature at this less visible level. An elegant fabric protector—a complete, well-executed solution—should probably not find its way into the environment, potentially contaminating water that people drink or food that they eat. It should stay where it is applied, and if it degrades, do so safely. Just as a life raft should not sink, and a Laserphaco Probe (described in the next chapter) should not blind, a fabric protector should not have the potential to cause any harm to humans. I am not suggesting, for one moment, that the inventors had any intention to cause harm, and when the potential risks were known, 3M acted to address the matter. However, this cannot mask the fact that the most creative solutions are harmonious and safe. For this reason, I have given this invention 2 out of 4 for elegance.

Genesis: One of the standout qualities of Scotchgard is the transferability that it embodies. This aspect of genesis describes the ability of some solutions to address different problems. In other words, the most creative solutions usually seem to have qualities beyond the problem they were designed to solve. This may be connected to the discovery-like process that gave rise to Scotchgard. I see this almost as a kind of *reverse divergent thinking*. Normally in creative problem-solving, we define the problem, and then we generate many possible solutions. In the case of Scotchgard, the solution was defined first, and then many possible problems were found for that solution. The underpinning cognitive process is the same, but the latter case may explain why some "inventions" score better for genesis. Once this water- and oil-repelling compound was created, Sherman, Smith, and 3M quickly developed a suite of problems, all of which make use of the basic qualities of the substance and a range of different situations in which these qualities are useful. Scotchgard scores well for genesis, at 3.5 out of 4, in recognition of its versatility as a solution to many problems.

Total: With a total of 12.5 out of 16, Scotchgard sits in the space between *high* and *very high* creativities. The weakness, in this case, is obvious. The solution is original, surprising, and radical in nature. It was, however, somewhat poorly executed, even if the properties that cause this weakness were not known at the time. The original Scotchgard, however, gives us an excellent opportunity to ask what would happen if these weaknesses were addressed. How would this change the creativity of the product? When 3M modified their product in 2003, they were attempting to address the weakness in execution that caused us to score the elegance of the product at 2. Does the *new* Scotchgard now deserve a higher score for creativity. To assess that, we must consider not only a likely increase in elegance, but whether or not this comes at any cost in effectiveness. This also forces us to decide what we value most in an invention: effectiveness, or novelty, or can these not be separated.

Chapter 12
The Digital Age (1981–Present): Complexity and Creativity

The final epoch in our study of creativity is the present. Although this spans the period from the 1980s to the present day, historians may eventually differentiate between the *early* Digital Age [running from the advent of the personal computer (PC) in 1981, to about 2015] and a *post-Digital Age*. In technological terms, the current part of the Digital Age—what I am suggesting may eventually be recognised separately—is characterised by a new Industrial Revolution, one that we have already discussed under the name *Industry 4.0*. The First Industrial Revolution is familiar to us from the Age of Enlightenment. The second Industrial Revolution is characterised by mass production, assembly lines and electricity, and corresponds approximately to the late Romantic period, and early Modern Age. The third Industrial Revolution is defined by the rise of computers and automation—intersecting with the late Modern Age and early Digital Age—but the *fourth* Industrial Revolution (Industry 4.0) represents the synthesis of data analytics, autonomous systems, and artificial intelligence in what is termed *cyber-physical systems*.

The first invention we will tackle in the Digital Age is another medical innovation. The *Laserphaco Probe* continues a long-standing theme of problems stemming from our fundamental need to live healthy lives. The Laserphaco Probe took advantage of related medical advances—for example, foldable artificial lenses—to deliver a step change in the invasiveness of surgery for cataracts. What was once untreatable, and then treatable only with a fairly difficult and invasive operation, has become almost routine thanks to this invention. The Laserphaco Probe also delivers disproportionately important outcomes to patients. People who might otherwise be blind as a result of cataracts have their vision restored with a short, simple operation, thanks to this innovation.

The last two innovations in our catalogue of *Femina Problematis Solvendis* should give us hope that many of the barriers that impeded female inventors in past eras are finally breaking down. Each of these final inventions came from the mind of a young woman, still at high school. The penultimate invention is the *Blissymbol printer*. This device converted the visual symbols used for communication by certain disabled individuals into text (and more recently, into synthesised speech). The final innovation in our series is the *thermoelectric flashlight*. Using only the heat generated

© Springer Nature Singapore Pte Ltd. 2020
D. H. Cropley, *Femina Problematis Solvendis—Problem solving Woman*,
https://doi.org/10.1007/978-981-15-3967-1_12

by a human hand, this device allows people living in countries with unreliable or incomplete electricity infrastructure to continue to work, study, and carry out other activities that need light. With this device, children can read and do their homework, for example, helping them to complete their education and raise their standard of living.

12.1   The Laserphaco Probe (1981)

The only thing worse than being blind is having sight but no vision—Helen Keller, American Author and Activist (1880–1968)

The first of our Digital Age inventions has a special significance for two reasons. The first is that its inventor, Dr. Patricia Bath, passed away while I was writing this book. Dr. Bath herself is a pioneer of more than just innovation. She was the first African-American female medical doctor to patent a medical invention, and she holds a number of other firsts in her field of ophthalmology in the USA. The second reason is that I have been the beneficiary of the expertise that Dr. Bath developed over her career, having had cataract surgery in both eyes. Many years ago, while visiting the Science Museum in London, I saw a diorama depicting a medieval scene involving the removal of a cataract from a person's eye. Back then, the process was very simple: hold patients down so they cannot move, insert a needle into the centre of their eye, and move the cloudy lens out of the way, all without anaesthetic! Although this left the patient without any lens, apparently it was better than no treatment at all. Having had cataract surgery on both eyes, as well as procedures to fix a detached retina, I can attest to the wonders that ophthalmologists now perform. Vale Dr. Bath!

Dr. Bath, in fact, was awarded a total of five patents for inventions all associated with the treatment of cataract—i.e. cloudy—lens. She achieved these over a period of 15 years, from 1988 to 2003, although we will concentrate on the first of these: the Laserphaco Probe. This is a tricky one to classify. I am tempted to focus on the energy that forms the basis of this device, but when we consider that the fundamental purpose is to remove a cloudy lens from a person's eye (and to do so safely and painlessly), then it becomes clear that the Laserphaco Probe is a material-handling system. Let's look more closely at this innovation.

12.1.1   What Was Invented?

The Laserphaco Probe that Patricia Bath first developed in 1981 is a surgical tool designed to vaporise cataracts by means of a laser. One of the critical features of this device is the fact that it allows the procedure of cataract removal to be conducted in a *minimally invasive* manner. Unlike previous techniques, ranging from the medieval method I described earlier, through to the more modern (lengthy extracapsular extraction) surgical techniques, minimally invasive procedures generally offer faster recovery, reduced pain, scarring and swelling, and greater accuracy.

Although Dr. Bath first conceived this innovation in 1981, it took a further seven years of refinement and development before she received her first patent in 1988. In fact, Dr. Bath holds a total of three patents related to the Laserphaco Probe. Like many medical innovations, where the effectiveness and safety of the device must be stringently verified before it is used with human patients, this device would only come into common use from the year 2000.

The history of cataract surgery is more complicated than just the removal of the cloudy lens. During the second half of the twentieth century, many parallel changes were taking place, beginning with the insertion of rigid replacement lenses. However, because of the size of these replacements—6–7 mm in size—there was little advantage to be gained from small incisions. A critical innovation that allowed the benefits and potential of the small-incision probes to be exploited was the development of *folding* lenses.

The Laserphaco Probe (see Fig. 12.1), through an incision of no more than 1 mm, uses the laser to dissolve, or *ablate*, the cloudy lens. The actual laser radiation is transmitted via a flexible optical fibre. This allows the laser energy to be applied more precisely to the affected lens. The optical fibre is partially surrounded by a sleeve that contains two sections. One part of the sleeve injects water to *irrigate* (i.e. flush) the treatment site, while the other side *aspirates* (i.e. sucks out) the water and ablated lens from eye. The particles created by the ablation process are so small that the probe, including the optical fibre and the irrigation/aspiration sleeve, is typically less than 1 mm in diameter.[1]

Dr. Patricia Bath would go on to receive two further patents, in 1998 and 1999, for refinements to the Laserphaco concept. She would also receive a patent, in 2000, for an ultrasonic cataract device, and finally, a patent in 2003 for a combined laser/ultrasound probe.

12.1.2 Why Was the Laserphaco Probe Invented?

The Laserphaco Probe solves a very clear problem: *how to remove cloudy lenses*. This invention illustrates an important concept, present in some of our innovations, but not others. In this case, the ablation and removal of cataracts is not really the goal. This is a necessary step in a larger problem, which is the question of *how to restore sight*. We could probably argue that every invention is, in some way, just a step in a broader problem-solving process. The quill pen, we decided, was designed to solve the problem *how to communicate in writing*, but even that is really just a step in solving a larger problem. One way to look at this is in relation to Maslow's hierarchy

---

[1] Image Credit: http://www.uspto.gov. Public Domain. Source: https://www.nlm.nih.gov/exhibition/aframsurgeons/morenotable.html.

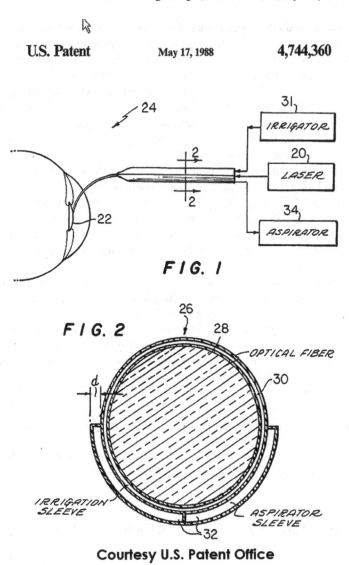

Fig. 12.1 Laserphaco Probe Patent, 1988

(see the Introduction). A given invention might solve a problem at a lower level of the hierarchy—for example, a physiological need—but together with other inventions, this contributes to satisfying higher-order needs (like the need for belongingness, esteem or self-actualisation). This seems to be the case with the Laserphaco Probe. It is really just a device for blasting eye lenses into tiny fragments. But when these are removed and replaced with a brand new, artificial lens, and a person's eyesight is restored, this has an impact far beyond the basic, physiological need to see.

12.1.3 How Creative Was the Laserphaco Probe?

Relevance and Effectiveness: Dr. Bath's invention could not have come into widespread use if it did not work. This is especially evident in medical inventions, where stringent approval processes govern the use of these devices in humans. You would probably therefore expect me to give the Laserphaco Probe a maximum score. However, our assessment of relevance and effectiveness is not just *correctness* and *performance*, but also the question of the invention fitting task constraints (i.e. *appropriateness*). What this means in practice is that the absolute best cataract removal (and lens replacement) procedure would involve no incision at all. A 1 mm incision is obviously a massive improvement over a 6 or 7 mm incision, but no incision at all would be the ultimate in minimally invasive surgery! For this reason, I give the Laserphaco Probe a score of 3.5 out of 4 for effectiveness.

Novelty: As we have seen elsewhere in this volume, an invention may be very strong in problematisation (how it helps to define the problem at hand), but slightly weaker in propulsion (how it sheds new light on the problem at hand). The Laserphaco Probe is a great example of this. It shows very clearly how and why previous solutions to the problem of cataract removal are deficient or unsatisfactory. It shows how to improve these deficiencies, and it also gets us thinking about how we could improve the solution even further. At the same time, it is not really a radical invention, but represents a clever and functional application of existing technologies. By using the ablative properties of lasers, the size, flexibility, and precision of optical fibres, and the modern capacity for precision engineering, an extremely small, highly functional tool could be created. It is, in other words, the integration of existing technologies, and incremental improvements, that give the Laserphaco Probe its qualities, rather than a radical departure from previous concepts. I give it 3.5 out of 4 for novelty.

Elegance: The stringent requirements for medical inventions are undoubtedly a driving force behind the elegance of Dr. Bath's device. It is well-executed and skilfully designed. It is compact, complete, and fully worked out, well-proportioned, and well-finished. It had to be, because anything less would be unacceptable in a device that is used for delicate and potentially dangerous medical procedures. There is also a pleasingness in the arrangement of the design shown in Fig. 11.3. The compactness of the probe, combining the optical fibre, irrigation, and aspiration sleeves, is clever and neat. It is an easy decision to give this invention a maximum of 4 for its elegance.

Genesis: Under the category of novelty, I deducted half a point, largely because it represented more incremental than radical qualities of innovation. As I indicated, it was a pinnacle of existing technologies. I have often explained that novelty is more about the invention itself, whereas genesis is more about the external impact of the invention, and this is the case here. Even though the Laserphaco Probe was not strongly radical, or paradigm-breaking in nature, its external impact is high. For example, it serves as a novel basis for further work. What other ablative technologies (aside from a laser) could be miniaturised and embedded in a tiny probe? Could an ablative technology be used that does not require irrigation or aspiration, thus allowing the design of an even smaller probe? Another important quality of genesis is also the new benchmark that this device established for the removal of cataracts. The strong genesis of the Laserphaco Probe lies in how it changes our thinking about the problem of treating cataracts. The fact that Dr. Bath herself developed and patented further approaches to this same problem is evidence of the genesis of this particular invention. I give it 4 out of 4 in this category.

Total: Dr. Patricia Bath's Laserphaco Probe scores an extremely strong, *very high* 15 out of 16 for creativity. This is one of the strongest in our catalogue of *Femina Problematis Solvendis* inventions. The only weaknesses that hold it back from a perfect score are a slight deficiency in effectiveness—although *minimally* invasive, it is, nevertheless, still invasive—and a slight weakness in novelty. This criticism is only that it is a novel *integration* of existing technologies and not a radical new technology in the sense, for example, of a previously unknown, non-invasive method by which cloudy lenses are dissolved and absorbed into the body, perhaps with some external radiation, or even a medicine taken by the patient. Take nothing away from Dr. Bath's invention, however, which, as I can attest, is a heck of a lot better than invasive treatments, or no treatment at all!

12.2 Blissymbol Printer (1984)

Mankind's greatest achievements have come about by talking, and its greatest failures by not talking—Stephen Hawking, English Theoretical Physicist (1942–2018)

The penultimate invention in our journey through the history of *Femina Problematis Solvendis* is remarkable because it was the twelve-year-old Canadian, Rachel Zimmerman Brachman, who developed the Blissymbol Printer in 1984.

One of the challenges for many people with severe physical disabilities is the ability to communicate. Think of the case of Professor Stephen Hawking. Motor Neurone Disease began to affect his voluntary muscle control from the age of 21, eventually robbing him of the ability even to speak. Cerebral Palsy is another disease that, in severe cases, impairs an individual's ability to speak and communicate. Tragically, this disease is the most common physical disability in childhood, often placing enormous physical constraints on children who are cognitively unimpaired. For these individuals, the ability to communicate is essential to their ability to develop, undertake education, and as adults to lead productive and fulfilling lives.

In the 1960s, Blissymbols—an ideographic writing system devised by Charles K. Bliss—became popular as a means for teaching disabled people to communicate. However, a limitation of this system remained the fact that it required the help of someone else. In other words, the disabled individual could not use the system independently. This is where the twelve-year-old Rachel Zimmerman[2] comes into the picture. We will now explore her information-handling system a little more closely.

---

[2] I will use her maiden name here, reflecting her name at the time she invented the Blissymbol printer.

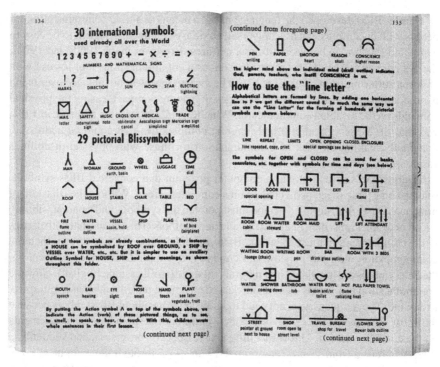

Fig. 12.2 Examples of Blissymbols

12.2.1 What Was Invented?

The Blissymbol printer (see Fig. 12.2 for an example) was developed to convert into written text the *ideograms*[3] conceived by Charles Bliss. Bliss devised a system of several hundred symbols, now about 900 in number, some of which are *ideographic* (abstract in nature), some of which are *pictographic* (literal representations of objects), and some of which are *composite* (two or more basic symbols are combined to create a new meaning). Together with grammar symbols (a small square, cone or "v"), which convert a basic symbol into, respectively, a *thing*, a *value*, or an *evaluation*, the Bliss system allows a fairly sophisticated and flexible means of communication independent of natural languages such as English, Arabic, and Chinese.[4]

---

[3]Ideograms are graphical symbols used to represent ideas or concepts. A simple example is a "no dogs allowed" sign consisting of a silhouette of a dog surrounded by a red circle, with a red diagonal line through the silhouette. The ideogram is independent of any specific language. Many ideograms are self-explanatory, while some are more abstract, and must be learned.

[4]This figure shows a selection of Blissymbols from the book Mr. Symbol Man (1976) by Charles K. Bliss and Shirley McNaughton. Published by Semantography-Blissymbolics, Sydney.

Charles Bliss developed the language not as a means of aiding communication for people with disabilities, but as a universal, auxiliary language, designed to be easy to learn, and as a bridge between different cultures. However, in the 1960s and 1970s, the system became popular, first in Canada, as a communications tool for use with children suffering from cerebral palsy.[5] The system has clear advantages over using natural language to spell out long messages. It is more compact, easing the physical demands on the disabled communicator. What Rachel Zimmerman realised, in 1984, was that the Bliss system still had inherent complexities that could be addressed, potentially making the task of communication for those with physical disabilities easier still.

Zimmerman's invention was a computer program that converted Bliss symbols, selected on a large touch pad, into text on a computer screen. Without the Blissymbol printer, a non-verbal, disabled person must point to the symbols, and a parent, sibling, friend or teacher must then interpret the selected symbols. This means that both parties in the communication must *speak* the Blissymbol language. If they do not, then the communication may still be slow and cumbersome. With Zimmerman's invention, the Blissymbols are translated into natural language (we'll assume English). Now the communication problem is significantly eased. A non-verbal, disabled user can *speak* to anyone. The recipient does not need to know anything about Blissymbols because the software converts the symbols and their grammar into plan language statements on the computer screen. Throw in a little text-to-speech conversion, and suddenly, a person suffering from severe cerebral palsy can *speak* through the computer! For a person with such a condition, this is more than just a matter of communication. The ability to communicate allows the fulfilment of higher-order needs for friendship, love, accomplishment, and self-actualisation.

12.2.2   Why Was the Blissymbol Printer Invented?

The Blissymbol printer is a very good example of a solution to a lower-level need (in the sense of Maslow) leading to the satisfaction of higher-order needs. When we are fighting for our next breath of air, all that matters is getting that next lungful. Once we have air to breathe, and water to drink, we can begin to concern ourselves with food, shelter, and warmth. We could argue that the ability to breathe, therefore, is at the heart of every problem. While that's probably true, it goes too far. However, the ability to communicate does tie directly to higher-order needs, as I mentioned earlier. Perhaps it is the case that lower-order needs are more independent, while higher-order needs are more integrated, more interconnected? Whatever the case, the Blissymbol printer transformed the ability of non-verbal sufferers of cerebral palsy to communicate. Many children with this condition are cognitively unimpaired. In

---

[5]Cerebral Palsy results from damage to the developing brain before, during or after birth. The condition typically affects muscle coordination and control. Severe cases may inhibit speech, and therefore communication.

blunt terms, they are intellectually normal, but trapped in a disabled body. They have the same intellectual needs as any able-bodied child. They want to go to school, learn new things, communicate their thoughts, desires, and dreams, and lead productive lives. The problem people with this condition face is *how to communicate*. This is the first instance where we have expressed our problem in only three words, without the noun!

12.2.3 How Creative Was the Blissymbol Printer?

Relevance and Effectiveness: Rachel Zimmerman's invention was an excellent integration of two mature technologies. Although it might not seem so to modern readers, software in 1984 was already fairly sophisticated. The computer power that we take for granted today may have been unavailable, but desktop machines were in common use, even in schools, and coding, albeit in languages like BASIC, was becoming a normal part of school curricula. Similarly, the *technology* of Blissymbols was well-established. What this meant was that combining these two things together was technically feasible. The actual performance of the Blissymbol printer, and its ability to satisfy reasonable constraints (e.g. to incorporate all available symbols and possible combinations), may have been somewhat suboptimal, but we should perhaps look at it as a concept demonstrator, or a sophisticated prototype (think back to the Antikythera Mechanism, or Hedy Lamarr's Radio-Controlled Torpedo). This *work-in-progress* feel to the first Blissymbol Printer means that I am giving it a score of 3.5 out of 4. It is tempting to cut Rachel Zimmerman some slack, given that she was only 12 at the time. On the other hand, her invention still performs strongly in comparison to those of both women and men who had many more years of education and experience to draw on.

Novelty: As we draw close to the end of this catalogue of *Femina Problematis Solvendis* inventions, there is one observation that I can make. There are more inventions in this book with a 4 for novelty than in the previous volume, *Homo Problematis Solvendis*. Whether that is because I am assessing the inventions differently (perhaps in part because of what I have learned as I've gone through this process), or because the inventions themselves are qualitatively different, is hard to say. Regardless, the Blissymbol printer cannot be faulted for its ability both to define the problem and to shed new light on the problem. It was a leap forward in the process of facilitating communication for non-verbal, disabled people. It was clearly a quantum leap forward in comparison to a non-computerised version of Blissymbol communication. It was more than just an incremental improvement and also opened up our understanding of how software and computers might be applied to problems associated with disability. The Blissymbol printer deserves a score of 4 out of 4 for novelty.

Elegance: This invention was reasonably well-executed. I always caution that we must not judge past inventions through a modern lens: that would be unfair. Thus, it would be unreasonable to criticise the inventor of the prehistoric shoe for not making use of composite materials! However, I think we can apply a critical eye to the execution of the Blissymbol printer, without placing impossible expectations on this innovation. As good as it was, it had some potential to be refined, not just in terms of its functionality (i.e. effectiveness) but in the quality and *slickness* of its execution. This probably reflects the *prototype* nature of the invention. Improvements would be made to the interface, to the layout, to the range of available symbols, and so on. At the same time, we can also appreciate an inventor sacrificing some of these more aesthetic qualities for the sake of getting a product into the market and into the hands of users. Sometimes, there is great benefit in the feedback obtained from *rapid prototyping*. This softens the score that we can give for elegance to 3 out of 4, but also shows the potential of this invention for further refinement and development.

Genesis: We see, once again, a typical pattern for the characteristics of creativity in this invention. Not only do effectiveness and elegance tend to follow each other, but those two elements may be a little weak, even though an invention is high in novelty and genesis. This seems to make intuitive sense. Invention is about new solutions that push boundaries. Newness is paramount. Of course, as I have often said, a solution must solve the problems it is meant to solve, but it can do this reasonably well, while still being a highly innovative idea. In other words, if you had a choice between giving up a little either for *effectiveness + elegance*, or for *novelty + genesis*, then an inventor might choose the former. This is what happens with concept demonstrators and prototypes. The idea is to get the *new* idea out into the hands of the user quickly, even if it is slightly imperfect. That seems to make more sense than getting a *perfect* solution into the market, only to find that you waited too long, and a competitor has launched a rival product. The Blissymbol printer generally scores well for genesis. It is a novel basis for further work on the same problem: exploiting the power of computers and software to tackle problems related to disability (and also communication more generally). *Google Translate* is a salient, modern example. I can now communicate with people who speak other languages, not by learning their language, but by letting my smartphone work as the translator. The Blissymbol printer also established a new benchmark for judging solutions to this problem, and generally established a new, computer-aided paradigm in this field. For these reasons, I have given it a score of 3.5 for genesis.

Total: That brings us to a score of 14 out of 16 for Rachel Zimmerman's invention. This places the Blissymbol printer well into the *very high* range for creativity. As I have already suggested, the room for improvement here lay in the completeness of the solution, its usability, and its execution. None of these things was bad, but each had room for further improvement and development. We also saw that this is a common dilemma faced by inventors. Do you spend time perfecting the invention, or is there greater value in getting the novel idea into the hands of users as quickly as possible?

12.3 The Thermoelectric Flashlight (2013)

Renewable energy is an essential part of our strategy of decarbonisation, decentralisation, as well as digitalisation of energy—Isabelle Kocher, French Business Executive, CEO of Engie (1966–)

We have made it to the final invention in the Digital Age! You might have expected something super-digital for this finale, and I do not blame you. We are now firmly into not just the Digital Age, but an era of *digital transformation*. I have already mentioned the Fourth Industrial Revolution, or Industry 4.0, in which we are beginning to see and experience the potential of many of the products of the digital age. Thus, as we move into the third decade of the twenty-first century, the impact of systems including AI, machine learning, robotics, and the Internet of Things is causing fundamental changes to the way we work, learn, and play. Notwithstanding this seismic shift in society, we finish our exploration of *Femina Problematis Solvendis* with something quite simple.

One thing that has been critical to centuries of human development, and is more important now than ever before, is energy. Without electrical energy, Industry 4.0 would stop dead in its tracks. How would we surf the web and communicate, let alone power the digital transformation of society, without this versatile form of energy? Yet, for many people,[6] the need for electrical energy is much simpler. Even in a world with one smart phone per head of population, there are still many people for whom light and heat are still much more basic concerns. The *thermoelectric flashlight* is an example of a simple energy-handling system that reminds us of a need that is as old as our species: the need for light!

12.3.1 What Was Invented?

Ann Makosinski, from the Canadian province of British Columbia, was motivated by a remarkably simple issue. Visiting her mother's native Philippines, she observed the fact that many high school students were unable to study at night because their homes lacked electricity. In fact, it is estimated that some 16 million people in that country do not have access to electricity. Compare this to countries like the USA or Australia, where 100% of the population enjoy the benefits of electricity.[7] Even though the overall rate in the Philippines is 93%, this is unevenly distributed between urban and rural populations and still leaves many people without something as simple as artificial light at night. In some countries (e.g. Chad, in Africa), the figure is as low as 10.9%. If you recall that the introduction of widespread gas lighting in England in the early nineteenth century was credited with raising rates of literacy in that country,

---

[6]See: https://www.gatesnotes.com/Climate-and-Energy.

[7]For recent (2017) estimates on the rates of electrification in different countries, take a look at the World Bank data available at this link: https://data.worldbank.org/indicator/eg.elc.accs.zs.

you can see that access to reliable sources of artificial light is about more than just convenience. It is fundamental to a society's development and to lifting people out of poverty.

Ann Makosinski recognised a clear need and set out to develop a solution. In 2013, while still in high school, Makosinski entered the Google Science Fair.[8] She understood, of course, that the human body generates heat. That heat is a by-product of the processes that convert the food we eat into the energy that powers our muscles, our brain, and keeps us alive.[9] Like any similar process—take a light bulb for example—the conversion process is never 100% efficient. An incandescent light bulb converts only about 10% of the incoming electrical energy into light with the rest being turned into heat. LED lights, on the other hand, are much more efficient. About 90% of the incoming electrical energy is converted into light. The human body is about 25% energy efficient. In other words, a substantial proportion of the energy we consume as food ends up as heat. What Makosinski realised was that this heat could be put to useful work, to power a flashlight.

Ann Makosinski knew about the *Thermoelectric Effect* from her physics lessons. This general term refers to the fact that when certain combinations of two dissimilar electrical conductors (e.g. copper and constantan wires) are brought into contact to form a *junction*, and when the unconnected ends of the conductors are at a different temperature to the junction, a voltage is generated. Thermocouples are widely used to measure temperature, but the generated voltage is also, of course, available for other purposes. Even though the voltage generated is small (typically in the region of microvolts, but possibly as high as tens of millivolts), it can still be put to good use.

Makosinski used thermoelectric devices known as *Peltier tiles* (Fig. 12.3) as the chief element of her design. Using a combination of four tiles, the heat from a human hand, and a hollow design that allowed the cool side of the tiles to remain at ambient temperature, she succeeded in creating a temperature difference of about 5 °C between the *hot* and *cold* sides of the thermoelectric components. This temperature difference was enough to power three LEDs, providing a useable source of light with no moving parts![10]

12.3.2   Why Was the Thermoelectric Flashlight Invented?

Ann Makosinski solved a clear problem. For millennia, humans have tried to find ways to light the darkness. Whether for safety or for comfort, the challenge of artificial

---

[8]The Google Science Fair is an online science competition, first established in 2011, sponsored by Google and other partners.

[9]The largest single consumers of energy in our bodies are actually our liver and our spleen. These account for about 27% of what is used when we are at rest. Our brains come next at about 19%.

[10]Image Credit: michbich. Source: https://commons.wikimedia.org/wiki/File:Peltierelement.png. Creative Commons 3.0: https://creativecommons.org/licenses/by-sa/3.0/legalcode.

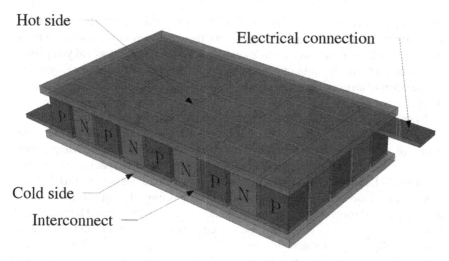

Fig. 12.3 Schematic of a Peltier device

sources of light has been a constant for humanity. Fire has always been a staple for this purpose, but it is often hard to control, needs sometimes costly or scarce resources, and generates significant waste products and copious amounts of heat that may or may not be desired. Candles and oil lamps addressed some of these problems but retained many of the limitations of fire. Even commercial electricity powering first incandescent, and later fluorescent and electronic lights (LEDs), has limitations, not least of which has been the pollution generated by coal-fired power stations. What Ann Makosinski developed was a truly versatile, portable, flexible, and clean solution to the problem of *how to generate light*.

12.3.3 How Creative Was the Thermoelectric Flashlight?

Relevance and Effectiveness: I have read some critiques of Ann Makosinski's thermoelectric flashlight. One of the chief concerns is the brightness of the device. This sort of criticism, however, seems to violate our rule about considering these inventions in their own right, judging them only as solutions to the identified problems. The point here is that we could shoot down any of the inventions in this book simply by arguing that, for example, there were other ways to secure a strap around your waist than a buckle, or different ways of writing than the quill pen. The point of our analyses of creativity is to look at the merits of the invention against our four criteria, understanding that something may be less effective than some alternatives, but infinitely more novel. Many of the inventions we have considered were superseded by other, better solutions, but they can still be analysed for their creativity at the time they were invented. My point is that we know there are brighter flashlights available.

But brightness is not the only consideration. If those brighter flashlights need expensive batteries, or access to domestic electricity, then they may not be a solution to the problem. If you were stranded on a desert island, which flashlight would you prefer? As a solution to the problem of generating light, with the constraint of *cheaply*, and *independent of any external power source*, Ann Makosinski's invention is a 4 out of 4 for effectiveness.

Novelty: Where we can subtract a little from the creativity score is against novelty. This is not a criticism of the device, but for me highlights the strongly incremental nature of the thermoelectric flashlight. Makosinski drew on long-standing principles of physics, proven electronic devices, and a tried-and-tested packaging concept. What she did, however, was find a new way to put these devices to use. Her invention is an evolution of the existing paradigm that has moved from portable candles and oil lamps, through battery operated torches, to solar-powered and hand-cranked flashlights, and now to a thermoelectric source of power. It is a splendid example of extending the known in a new direction, without breaking out of an old paradigm, or establishing a radically new paradigm. Incrementation—making things better, faster, and cheaper—is the bread and butter of innovative engineering, and this invention is a fitting example. I give it a 3 out of 4 in this category.

Elegance: Like many of our inventions, elegance and effectiveness go hand-in-hand. This case is no exception. In fact, the thermoelectric flashlight epitomises the benefits of strong elegance for strong effectiveness. The device had to be well-executed to achieve the level of performance that it did. Without careful attention to the physical arrangement of the components, best seen in the creation of a hollow interior to help set up the necessary temperature differential, the device may not have worked. The invention also needs to be designed with the end-user in mind, to ensure that it meets the underlying need for supplying useable light. It is well-finished, attractively designed and skilfully executed, scoring 4 out of 4 for elegance.

Genesis: Thus, we reach our final category. The thermoelectric flashlight has reasonably strong elements of genesis. To some degree, it changes how we think of the problem. Conventional flashlight designers might never consider constraints such as access to batteries. The approach is rather like the design of racing cars: see how far you can go in terms of brightness and robustness. Make a torch that is so bright that it can illuminate things 400 m away. Make it out of titanium so that it can double as a hammer! But what if this is not really the problem? What if the problem is not just ever-increasing brightness? Is the problem that a flashlight solves to be as bright as possible, or as portable as possible, or as reliable as possible or something else? Ann Makosinski's design re-interpreted the question and found that a different answer might be what was needed. Her answer was not overwhelmingly radical in nature, but it suggested a new way of looking at the problem (germinality) and produced an incrementally new, and somewhat different solution. I finish with a genesis score of 3 out of 4 for her invention.

Total: The thermoelectric flashlight completes our catalogue with a total score of 14 out of 16 for creativity. Sitting well inside the *very high* range, this invention is an excellent example of how some innovations extend and expand on the existing paradigm, squeezing more value out of an idea, and looking to refine and improve solutions as much as possible. It is also an interesting example of an invention that had a lot of people asking themselves *why didn't I think of that*? What it really seems to illustrate is that there is no limit, and no barrier, to creativity, if we keep open minds.

Chapter 13
The Innovation Scoreboard

We have reached the end of our exploration of thirty-one inventions by women, across ten key periods of human history. However, the story does not end there, because we now need to conduct some analysis. In the companion to this book—*Homo Problematis Solvendis*—I examined that set of inventions, and their creativity scores, to see if there were any discernible patterns that we could extract. In this book, we have the opportunity not only to survey our new set of inventions, but also to compare these with the previous set. I should hasten to add that my underlying purpose is *not* driven by any expectation that we will see major differences between men and women. Rather, I am keen to see if we find any further evidence of differences between eras, or between the broad classifications of invention (energy-, information-, and material-handling systems). Having said that, it will be interesting to see if there is any noticeable difference between the creativity of inventions by the different genders! If such a difference does exist, keep in mind the fragility of our sample. A total of 61 inventions, scored only by me, is not a very robust sample. Again, if such a difference exists, my guess is that it will be found not in absolute levels of creativity, but in the detail. Is it possible that the inventions by women are, for example, generally higher in novelty or elegance, compared with those by men? Or, is the set of *Femina* innovations generally more effective than those developed by men?

We'll begin with the *Femina* inventions, before making some comparisons with the male inventions of *Homo Problematis Solvendis*. The following list shows the *Femina* inventions ranked from highest to lowest on the *total creativity* score (out of 16). There does not seem to be any pattern here regarding the era, but we'll look more closely at the different time periods, and types of invention shortly (i.e. energy-, material-, and information-handling).

1. Kevlar (16)
2. Astronomical Tables (16)
3. The Compiler (15.5)
4. The Propeller (15)
5. The Laserphaco Probe (15)

© Springer Nature Singapore Pte Ltd. 2020
D. H. Cropley, *Femina Problematis Solvendis—Problem solving Woman*,
https://doi.org/10.1007/978-981-15-3967-1_13

6. The Buckle (15)
7. The Computer Program (14.5)
8. The Radio-Controlled Torpedo (14.5)
9. The Wheel (14)
10. The Verge Escapement (14)
11. The Corn Mill (14)
12. Coade Stone (14)
13. Straw Weaving (14)
14. The Berkeley Horse (14)
15. The Waterproof Diaper (14)
16. The Blissymbol Printer (14)
17. The Thermoelectric Flashlight (14)
18. Soap (13)
19. The Bain-Marie (13)
20. The Quill Pen (13)
21. Feminism (13)
22. The Two-Handed Internal Rotation (13)
23. The Brassiere (13)
24. Shoes (12.5)
25. The Antikythera Mechanism (12.5)
26. Lingua Ignota (12.5)
27. The Toothbrush (12.5)
28. Silk Design (12.5)
29. Scotchgard (12.5)
30. The Life Raft (12)
31. Palliative Medicine (11).

We can gain some additional insights into the creativity of these inventions by exploring some of the subcategories of creativity used in our assessment scale. I selected inventions first and foremost because they should be *good solutions* to a problem. This means that we would expect to see a lot of 4's for *effectiveness*, and that is certainly the case (see Appendix B for the full breakdown of effectiveness, novelty, and so on). It is somewhat more revealing to look at the combination of *effectiveness* + *elegance* (with a maximum score of 8) to get a true picture of the *goodness* of solutions. I like to think of these two criteria together as defining *functionally* creative solutions, i.e. those that do the job they are designed to, and that are well-executed. Within our catalogue of *Femina* inventions, we find eight with a maximum *functional* score:

1. The Buckle (8);
2. Lingua Ignota (8);
3. Astronomical Tables (8);
4. Straw Weaving (8);
5. The Berkeley Horse (8);
6. The Propeller (8);

7. Kevlar (8);
8. The Thermoelectric Flashlight (8).

These *highly functional* solutions represent a fairly even spread across time periods, ranging from 700 BCE (the Buckle), right up to the Thermoelectric Flashlight (in the Digital Age). In terms of the type of system, there is a modest bias to energy-handling systems (four solutions), with two each for material- and information-handling.

At the other end of the *functional* spectrum, we have the following low scores (out of 8):

1. Soap (5);
2. Feminism (5);
3. Palliative Medicine (5);
4. Scotchgard (5).

The *low functional* (keeping in mind, that is a relative term) solutions cluster somewhat in the period before 1600 CE (Scotchgard being the exception) and are evenly spread across the type of system.

It is also interesting to look at the top innovations by novelty—*the surprisers*—to see if any pattern or features stand out in this respect. The top *surprisers*, out of a maximum score of 4, are:

1. Soap (4);
2. Feminism (4);
3. Astronomical Tables (4);
4. The Corn Mill (4);
5. The Brassiere (4);
6. The Radio-Controlled Torpedo (4);
7. The Compiler (4);
8. Kevlar (4);
9. Scotchgard (4);
10. The Blissymbol Printer (4).

Here we have quite a long list of innovations spanning energy-, material-, and information-handling systems, as well as a good distribution across eras. One note of interest is that all three inventions from the Modern Age (1880–1950) scored the maximum for novelty.

At the lower end of the novelty scale, we find the following *least surprising* inventions:

1. Lingua Ignota (2.5);
2. Silk Design (2.5);
3. Straw Weaving (2.5).

No obvious pattern here except that the latter two are both from the Age of Enlightenment and are both material-handling solutions.

Looking only at the criterion of elegance—the *pleasing solutions*—we find a large number of solutions. Fourteen score the maximum for elegance (see Appendix B), with a fair distribution across eras and types of system. This itself is noteworthy when you consider that in *Homo Problematis Solvendis*, only five of the inventions scored a maximum for elegance. Are *Femina Problematis Solvendis* inventions generally better executed?

Look also at the *least pleasing* solutions, we find the following:

1. Soap (2);
2. Scotchgard (2);
3. The Bain-marie (2.5);
4. The Corn Mill (2.5).

No obvious pattern here, except that there are no information-handling solutions that appear as *least pleasing* solutions.

Finally, it is also interesting to look at the top inventions by *genesis*: what we can call the *disruptors* or *paradigm breakers*. In this category, out of a maximum score of 4, we find:

1. Soap (4);
2. Feminism (4);
3. Astronomical Tables (4);
4. The Corn Mill (4);
5. The Radio-Controlled Torpedo (4);
6. The Compiler (4);
7. Kevlar (4);
8. The Laserphaco Probe (4).

The most noticeable feature of this list is the close correlation with novelty. In the same way that effectiveness and elegance tend to occur together, so it appears that novelty and genesis are strongly linked. It is also interesting that only one energy-handling system appears on this list. At the other end of the paradigm-breaking spectrum, we have the following:

1. Lingua Ignota (2);
2. The Toothbrush (2.5);
3. Silk Design (2.5).

No clear pattern here except that each of these inventions was *highly functional*. Is genesis inversely proportional to functionality, at least to some degree?

Before we try to draw any over-arching conclusions, there are two more sets of scores to consider. First, how do the inventions compare when we look across the categories of energy-handling, material-handling and information-handling systems. Although we have slightly different numbers in these categories, the following average scores for total creativity (out of 16) are:

- Energy-handling systems (10 in total) = 13.70;
- Material-handling systems (12 in total) = 13.42;
- Information-handling systems (9 in total) = 13.94.

It is difficult to draw any firm conclusions from this. The differences in average scores are small—no more than half a point—and without a larger sample, and without more independent scorers (i.e. people other than me), it is risky to conclude that there is any significant difference in creativity on the basis of the type of system. Are information-handling systems inherently more creative than energy- and material-handling systems, or is that just a reflection of the inventions chosen, and even my personal biases and preferences?

Finally, we can also look at creativity scores in different time periods, and we find the following average scores (again, out of 16):

- Prehistory (3) = 13.17;
- The Classical Period (3) = 13.50;
- The Dark Ages (3) = 12.83;
- The Renaissance (3) = 12.50;
- The Age of Exploration (3) = 14.33;
- The Age of Enlightenment (3) = 13.50;
- The Romantic Period (4) = 13.88;
- The Modern Age (3) = 14.33;
- The Space Age (3) = 14.17;
- The Digital Age (3) = 14.33.

Although I am also reluctant to declare any firm conclusions from these figures, it appears, from our limited sample, that we have something of a clear trend in terms of the different time periods. Aside from a bulge around the period from 1650 through 1715 in the Age of Exploration, the general trend of the scores seems to lean towards the more modern time periods. There is a gentle upward progression in creativity scores as we move from prehistory through to the Digital Age.

So, how do our two groups compare? Are there any obvious or significant differences between the *Femina Problematis Solvendis* scores and the *Homo Problematis Solvendis* scores of the previous book? Working through the same categories, beginning with overall creativity, we can note that the overall average total creativity for the *Femina* inventions is 13.66, compared with 13.08 for the male inventions (see Table 13.1). In strict, mathematical terms, that just fails to show statistical significance. In other words, there is a roughly 7% chance that the difference in mean total scores is just a random effect. Conversely, there is a 93% chance, if you trust my scoring, that the difference in mean creativity is real. If we drill down to the individual components of creativity (effectiveness, novelty, elegance and genesis), we find the following.

Once again, considering that these scores were allocated only by me, and therefore could be biased and subject to error and random variation, there appears to be some sort of pattern to the differences between *Femina* and *Homo Problematis Solvendis*. The higher average total of the *Femina* inventions is not uniform across the four criteria of creativity, with the male inventions indicating slightly higher relevance and effectiveness, but weaker scores across the other three criteria.

Table 13.1 Mean creativity scores—*Femina* versus *Homo*

Criterion	*Femina P S*	*Homo P S*
Relevance and effectiveness	3.48	**3.72**
Novelty	**3.44**	3.20
Elegance	**3.40**	3.18
Genesis	**3.34**	2.97
Total	**13.66**	13.08

Table 13.2 Type of system and creativity—*Femina* versus *Homo*

Types of Systems	*Femina P S*	*Homo P S*
Energy-handling	**13.70**	13.43
Material-handling	**13.42**	12.50
Information-handling	**13.94**	13.00

Table 13.3 Era and creativity—*Femina* versus *Homo*

Era	*Femina P S*	*Homo P S*
Prehistory	**13.17**	12.83
The Classical Period	**13.50**	11.83
The Dark Ages	12.83	**13.67**
The Renaissance	**12.50**	12.33
The Age of Exploration	**14.33**	14.00
The Age of Enlightenment	13.50	**13.67**
The Romantic Period	**13.88**	13.67
The Modern Age	**14.33**	13.17
The Space Age	**14.17**	12.50
The Digital Age	**14.33**	13.17

Peering deeper into the types of system (Table 13.2), we can see that each category favours the *Femina* inventions. I am not sure what to conclude from this, except that the *Femina* scores are more evenly distributed across the three types.

Last, but not least, Table 13.3 shows the differences in *Femina* and *Homo Problematis Solvendis* inventions across the ten time periods.

As always, allowing for the fact that this is a small sample, and without the benefit of a more diverse range of judges, I would venture to suggest one noteworthy trend in this table. The male scores tended to jump around from one era to the next, sometimes higher, sometimes lower. The female scores, however, show more of a steady increase in creativity across the eras. Does this reflect changes in the conditions of invention that have affected men and women differently throughout history? In particular, is this some sort of evidence that the barriers and challenges that female inventors have faced are generally improving?

As we draw our study of *Femina Problematis Solvendis* to a close, there are many questions remaining. Both this book, and the predecessor, have only really scratched the surface of the history of creativity and innovation. Creativity is a complex competency, combining personal dispositions, attitudes and knowledge, with special thinking skills, all affected by the environment in which the inventor operates. One of the best ways to study creativity is to look at the creative outputs that people produce, and by assessing these inventions for their effectiveness, novelty, elegance, and genesis, I hope that you have gained some insight into how this complex system works. In particular, I hope that I have shed some greater light on the history of creative, problem-solving women, the challenges they have faced, and their considerable achievements.

I will close with a quote from British–German physician and psychotherapist Charlotte Wolff (1904–1986). She said: *Women have always been the guardians of wisdom and humanity which makes them natural, but usually secret, rulers. The time has come for them to rule openly, but together with and not against men.* If we are to solve the challenges that we face in the twenty-first century, then it is time to make sure that we draw on the creative problem-solving skills of *every member* of our clever, inventive, and wise species.

Appendix A
The Creative Solution Diagnosis Scale (CSDS)

The Creative Solution Diagnosis Scale (CSDS)[1] is a measurement scale designed to help people assess the creativity of artefacts. These artefacts can be any outcome of an activity: for example, written essays, poems, constructions, artefacts. Equally, the artefact may be any solution to a problem.

The CSDS is designed to assist in the process of understanding what makes something (a solution, or *product*) creative. A key concept from research is that the creativity of solutions is not just defined by something that is new, or different, but also something that is effective and relevant to the problem or task at hand.

We also know that it can be hard for people to recognise and judge creativity. Experts are pretty good at doing this in their field of expertise, but what about other people, e.g. outside a given field? The CSDS is designed to help anybody judge the creativity of anything, and to do so in a rigorous and systematic fashion. In this way, the CSDS can also be used, for example, by teachers as a means of giving students formative feedback on their work, when creativity is a focus.

To do this, the CSDS uses a number of indicators grouped into five categories.

The five main categories that define the creativity of an artefact are:

- Relevance and Effectiveness—*the artefact is fit for purpose*;
- Problematisation—*the artefact helps to define the problem/task at hand*;
- Propulsion—*the artefact sheds new light on the problem or task*;
- Elegance—*the artefact is well-executed*;
- Genesis—*the artefact changes how the problem/task is understood*.

However, these can still be hard for people to recognise in an artefact without some further guidance. Therefore, we have developed 21 more detailed *indicators* of creativity—these are easily recognisable characteristics of an artefact that help define creativity. For example:

- Performance = "the artefact does what it is supposed to do".

---

[1]See also, Cropley, D. H. and Cropley, A. J. (2016). Promoting creativity through assessment: A formative CAA tool for teachers, *Educational Technology Magazine,* 56:6, pp. 17–24.

© Springer Nature Singapore Pte Ltd. 2020
D. H. Cropley, *Femina Problematis Solvendis—Problem solving Woman*,
https://doi.org/10.1007/978-981-15-3967-1

To make an assessment of the creativity of an artefact a person goes through the 21 indicators and rates the artefact on a scale from "0" (that indicator does not apply to the artefact), through to "4" (that indicator applies very much to the artefact). By combining these ratings together, we get a very detailed assessment of the artefact's creativity, and one that can be used for formative assessment—to help students understand what to do to increase the creativity of the things that they produce.

The full CSDS assessment scale is shown below:

A. **Relevance and Effectiveness** (the artefact is fit for purpose):

   a. CORRECTNESS (the artefact accurately reflects conventional knowledge and/or techniques);
   b. PERFORMANCE (the artefact does what it is supposed to do);
   c. APPROPRIATENESS (the artefact fits within task constraints).

B. **Problematisation** (the artefact helps to define the problem/task at hand):

   a. DIAGNOSIS (the artefact draws attention to shortcomings in other existing artefacts);
   b. PRESCRIPTION (the artefact shows how existing artefacts could be improved);
   c. PROGNOSIS (the artefact helps the observer to anticipate likely effects of changes).

C. **Propulsion** (the artefact sheds new light on the problem/task):

   a. REDIRECTION (the artefact shows how to extend the known in a new direction);
   b. REINITIATION (the artefact indicates a radically new approach);
   c. REDEFINITION (the artefact helps the observer see new and different ways of using the artefact);
   d. GENERATION (the artefact offers a fundamentally new perspective on possible artefacts).

D. **Elegance** (the artefact is well-executed):

   a. CONVINCINGNESS (the observer sees the artefact as skilfully executed, well-finished);
   b. PLEASINGNESS (the observer finds the artefact neat, well done);
   c. COMPLETENESS (the artefact is well worked out and "rounded");
   d. GRACEFULNESS (the artefact is well-proportioned, nicely formed);
   e. HARMONIOUSNESS (the elements of the artefact fit together in a consistent way).

E. **Genesis** (the artefact changes how the problem/task is understood):

   a. FOUNDATIONALITY (the artefact suggests a novel basis for further work);
   b. TRANSFERABILITY (the artefact offers ideas for solving apparently unrelated problems);

    c. GERMINALITY (the artefact suggests new ways of looking at existing problems);

    d. SEMINALITY (the artefact draws attention to previously unnoticed problems);

    e. VISION (the artefact suggests new norms for judging other artefacts-existing or new);

    f. PATHFINDING (the artefact opens up a new conceptualisation of the issues).

In this book, readers will note that I have used only four criteria to assess each invention. The only difference in the scales is that I frequently combine *problematisation* and *propulsion* together as a single criterion of *novelty*. Aside from that, in making my assessments of creativity for each invention, I used the 21 indicators listed, and calculated a single score, to the nearest half point, for each of the four main criteria.

Appendix B
Tables of Creativity Scores

The following table (Table B.1) lists the thirty-one inventions of *Femina Problematis*

Table B.1 Creativity scores—*Femina Problematis Solvendis*

Invention (Type)	Effectiveness	Novelty	Elegance	Genesis	Total
Shoes (E)	3.5	3	3	3	12.5
The Wheel (E)	4	3.5	3	3.5	14
Soap (M)	3	4	2	4	13
The Buckle (M)	4	3.5	4	3.5	15
The Antikythera Mechanism (I)	2.5	3	4	3	12.5
The Bain-marie (E)	4	3.5	2.5	3	13
Lingua Ignota (I)	4	2.5	4	2	12.5
The Quill Pen (I)	3.5	3	3.5	3	13
Feminism (I)	2	4	3	4	13
The Toothbrush (M)	3	3	4	2.5	12.5
Palliative Medicine (M)	2.5	3	2.5	3	11
The Verge Escapement (E)	3.5	3.5	3.5	3.5	14
The Two-Handed Internal Rotation (M)	3.5	3.5	3	3	13
Astronomical Tables (I)	4	4	4	4	16
The Corn Mill (M)	3.5	4	2.5	4	14
Coade Stone (M)	4	3.5	3.5	3	14
Silk Design (M)	3.5	2.5	4	2.5	12.5
Straw Weaving (M)	4	2.5	4	3.5	14
The Berkeley Horse (E)	4	3	4	3	14
The Computer Program (I)	3.5	3.5	4	3.5	14.5
The Propeller (E)	4	3.5	4	3.5	15
The Life Raft (E)	3	3	3	3	12
The Brassiere (M)	3	4	3	3	13
Radio-Controlled Torpedo (I)	3.5	4	3	4	14.5
The Compiler (I)	3.5	4	4	4	15.5
The Waterproof Diaper (M)	3.5	3.5	3.5	3.5	14
Kevlar (E)	4	4	4	4	16
Scotchgard (E)	3	4	2	3.5	12.5
The Laserphaco Probe (M)	3.5	3.5	4	4	15
The Blissymbol Printer (I)	3.5	4	3	3.5	14
The Thermoelectric Flashlight (E)	4	3	4	3	14

© Springer Nature Singapore Pte Ltd. 2020
D. H. Cropley, *Femina Problematis Solvendis—Problem solving Woman*,
https://doi.org/10.1007/978-981-15-3967-1

Solvendis in chronological order (and system type: E for energy-handling, and so on), with the details of their creativity scores, broken down by creativity category.

For reference, and for the purpose of comparing both male and female inventions, Table B.2 shows the *Homo Problematis Solvendis* inventions from the book previous to this one (shaded), and the current *Femina* inventions, all grouped according to time period.

Table B.2 Creativity scores for the sixty-one inventions (*Homo* and *Femina*)

Invention	Effectiveness	Novelty	Elegance	Genesis	Total
The Hand Axe	4	3	2.5	2	11.5
Shoes	3.5	3	3	3	12.5
Oars	4	3.5	3	2.5	13
The Wheel	4	3.5	3	3.5	14
Cuneiform	3.5	3.5	3.5	3.5	14
Soap	3	4	2	4	13
Standardised Coinage	3	2.5	2.5	3	11
The Buckle	4	3.5	4	3.5	15
The Construction Crane	3.5	3	3	3	12.5
The Antikythera Mechanism	2.5	3	4	3	12.5
The Julian Calendar	3.5	3	3	2.5	12
The Bain-marie	4	3.5	2.5	3	13
Paper	4	3.5	3	2.5	13
Lingua Ignota	4	2.5	4	2	12.5
The Spinning Wheel	4	3	3	3	13
The Quill Pen	3.5	3	3.5	3	13
The Crankshaft	4	3.5	3.5	4	15
Feminism	2	4	3	4	13
The Scythe	3	2.5	3	2.5	11
The Toothbrush	3	3	4	2.5	12.5
The Printing Press	4	3	3	2.5	12.5
Palliative Medicine	2.5	3	2.5	3	11
The Condom	4	3.5	3	3	13.5
The Verge Escapement	3.5	3.5	3.5	3.5	14
The Slide Rule	4	3	4	3	14
Astronomical Tables	4	4	4	4	16
The Pendulum Clock	3.5	4	4	3	14.5
Two-Handed Internal Rotation	3.5	3.5	3	3	13
The Steam Pump	3.5	3.5	2.5	4	13.5
The Corn Mill	3.5	4	2.5	4	14
The Leyden Jar	3.5	3.5	3	3.5	13.5
Silk Design	3.5	2.5	4	2.5	12.5
The Industrial Revolution	4	4	3.5	4	15.5
Coade Stone	4	3.5	3.5	3	14
The Smallpox Vaccination	3.5	3	2.5	3	12
Straw Weaving	4	2.5	4	3.5	14
The Velocipede	4	4	4	4	16
The Berkeley Horse	4	3	4	3	14
The Computer Program	3.5	3.5	4	3.5	14.5
Sewerage Systems	4	3	3.5	2	12.5
The Propeller	4	3.5	4	3.5	15
The Electric Lightbulb	4	3	3	2.5	12.5
The Life Raft	3	3	3	3	12
Steel-Girder Skyscrapers	4	3.5	4	3	14.5
The Brassiere	3	4	3	3	13
The Wright Flyer	3	3	3	2.5	11.5
The Radio-Controlled Torpedo	3.5	4	3	4	14.5
Antibiotics	4	3.5	3	3	13.5
The Compiler	3.5	4	4	4	15.5

(continued)

Table B.2 (continued)

Sputnik I	4	2.5	3.5	2.5	12.5
Waterproof Diaper	3.5	3.5	3.5	3.5	14
Nuclear Power	4	2.5	3.5	2.5	13
Kevlar	4	4	4	4	16
Antisense (Gene) Therapy	2.5	3.5	2.5	3.5	12
Scotchgard	3	4	2	3.5	12.5
The WWW	4	3	4	2	13
Laserphaco Probe	3.5	3.5	4	4	15
Artificial Skin	3.5	3	3	3.5	13
The Blissymbol Printer	3.5	4	3	3.5	14
Smart Phones	4	3	3	3.5	13.5
The Thermoelectric Flashlight	4	3	4	3	14

Bibliography

Bernstein, P. L. (1996). *Against the Gods: The Remarkable Story of Risk*. New York, NY: John Wiley and Sons Inc.

Brandt, A., & Eagleman, D. (2017). *The Runaway Species: How Human Creativity Remakes The World*. Edinburgh, UK: Canongate Books Ltd.

Cadbury, D. (2012). *Seven Wonders of the Industrial World*. London, UK: Harper Perennial.

Cropley, D. H. (2019). *Homo Problematis Solvendis—Problem Solving Man: A History of Human Creativity*. Singapore: Springer Nature.

Diamond, J. (1999). *Guns, Germs, and Steel: The fates of Human Societies*. New York, NY: W. W. Norton & Company Ltd.

Dickson, P. (2001). *Sputnik: The Shock of the Century*. London, UK: Walker Publishing Company.

Gertner, J. (2012). *The Idea Factory: Bell Labs and the Great Age of American Innovation*. London, UK: The Penguin Press.

Harari, Y. N. (2011). *Sapiens: A Brief History of Humankind*. London, UK: Vintage Books.

Holmes, R. (2010). *The Age of Wonder*. New York, NY: Vintage Books.

Khan, A. (2017). *Adapt: How we can Learn From Nature's Strangest Inventions*. London, UK: Atlantic Books.

Kirby, R. S., Withington, S., Darling, A. B., & Kilgour, F. G. (1990). *Engineering in History*. New York, NY: Dover Publications Inc.

Moran, S., Cropley, D. H., & Kaufman, J. C. (Eds.). (2014). *The Ethics of Creativity*. Basingstoke: Palgrave MacMillan.

Petroski, H. (1996). *Invention by Design: How Engineers get From Thought to Thing*. Cambridge, MA: Harvard University Press.

Schilling, M. A. (2018). *Quirky*. New York, NY: PublicAffairs.

© Springer Nature Singapore Pte Ltd. 2020
D. H. Cropley, *Femina Problematis Solvendis—Problem solving Woman*,
https://doi.org/10.1007/978-981-15-3967-1